Technically a Blog

Nicole Tietz-Sokolskaya

First edition 2024

Cover design by Vladimir Kašiković.

ISBN 979-8-9899299-0-0

Published by May 11 Publishing LLC
www.may11publishing.com

To Eugenia, who supported me through finding myself.
And to Sophia and Alexei, who are the loves of my life.

Table of contents

Table of contents

3

1 Introduction

I started my blog in 2015 with the header "Technically a Blog." It was a throwaway name that has come to capture my writing. Yes, it's technically a blog; and it's a blog about technical things. But it's so much more than that.

Now, it has become a place where I think and share ideas with the world. It's a place where I write to make a difference. And it's a place where my writing has affected people in significant ways.

I did not envision any of this when I started. My first posts were coming from a place of insecurity and arrogance mixed together in a hearty cocktail. I had some good ideas, but I was not a great writer, and I wanted popularity more than I cared about the craft.

That changed in 2022, which was a very catalytic year for me. During my 12-week batch at the Recurse Center, I vowed to write a post each week—and ended up writing 35,000 words. Writing on a schedule forced me to let go of grandiose ideas and get practical, and it gave me the practice of writing regularly that I needed to improve as a writer.

This volume includes first all my writing from 2022, followed by the back catalog from 2015 through 2021. The older posts have some gems in them, but they also show the naivete of a new grad who has no experience in the world.

There are inconsistencies throughout in title capitalization, because I hadn't settled into a style until 2023. There are also spelling mistakes. I've largely preserved these to better capture the evolution of my writing and style over time, where they're not very distracting.

I am so grateful that you've picked up this book and decided to spend some time reading things I've written. Please read with an open heart, and always feel free to reach out to me. I'd love to hear from you.

-Nicole ❤

2 Running an Effective Book Club at Work

Originally published on 2022.07.09.

Even with the wealth of information on web sites and in videos, books remain a great resource for learning. And they're great for group learning, too! We've run a book club at work[1] a few times. Some sessions were more successful than others.

The main way our book clubs faltered or failed was through severe **drop-off**. This is a proxy for a lot of things (losing interest, too time consuming, etc.) and is measurable. Some amount of drop-off is normal. But if you lose the majority of your club, something has gone wrong. We need to make sure we work on keeping attendance high!

Here are a eight things I've learned about how to make an at-work book club successful! These helped us keep attendance high and helped us all get a lot out of the books we read.

Pick a relevant book. When you're doing a book club at work, taking work time for it, this is kind of a given. I wouldn't run a book club on Haskell at my day job. Not because Haskell isn't great, but because it's not *relevant* for what we do at my employer. We've run book clubs on Python (our primary backend language), distributed systems, and machine learning. Each of these are critical to what we do, so we had lots of initial interest in each.

Pick an interesting book. Related to books being relevant, they also need to be interesting. A relevant book will get people to check out what the book is about. An interesting book will get them to join and keep attending. Ultimately, what's interesting is subjective. We've had good luck finding interesting books by polling coworkers for book suggestions. You start from books that people already want to read, and you can poll them to measure how many are interested in it!

Set expectations upfront. I like to make the first session just an info session. When we tried to have the first session cover a chapter, it was too much. It can be intimidating to have to read before you know what the format is or what the expectations are! And it's also just plain confusing. For the first session, just lay out what the expectations are: how often do you meet, what participation requires, and all that.

---

[1] https://remesh.blog

Make sure people can get the book. Ideally, the company should buy it for everyone. I know software engineers are generally well paid, but we all have different situations. Even just the inconvenience of buying the book can sometimes drive people off, or they delay too long and miss the window to start participating. The easiest way to get high initial participation is have work buy the book for everyone.

Make participation easy. If you make everyone prepare for every session, you'll lose folks. Quickly. We're all busy with our day jobs and it needs to be a small commitment to join. This comes down to how you structure the sessions. We like to run recap sessions: One person presents the chapter, then we discuss it after that. The key element here is that even if you don't read the chapter, attendance is still worthwhile because you can hear the recap and get some knowledge. If you focus on just discussion, you drive people out if they miss one week, and this leads to steep drop-off.

Set a fast schedule and follow through. Don't slip the schedule. Books are long, and you have to keep the momentum up to keep people reading. My preference is a weekly schedule with a significant chunk of reading each week. This can be difficult with all the other things in life, but it means you get done quicker, too.

Limit the length. In my experience, 10 weeks is about the limit for running a book club. After this, people will stop participating and get fatigued of it. When you're reading a denser book, like Designing Data-Intensive Applications[2], consider reading *part* of the book together and leave the rest for independent reading. We read parts 1 and 2 together and left part 3 for anyone who wanted to continue independently.

Rotate presenters. We do a recap-plus-discussion format, and we rotate who presents each chapter. If you have one person present each week, you lose out on people learning from presenting. You also make the program less sustainable: You cannot, as one person, sustain this for a long time and a lot of different books. Rotating presenters each week will make the program sustainable and allow it to continue for more than one book. And have a backup plan in case someone backs out, so you don't have to slip the schedule.

Running a book club is highly rewarding. You get to help everyone in the group learn a lot and bond together, and you develop new skills as the facilitator of the group. I hope this is helpful if you choose to run one at work. Keep in mind that there are many ways to run a successful book club, and these are just things that I found effective in a particular group.

---

[2]https://dataintensive.net/

If you have anything else that you've found highly effective, please reach out and let me know!

Thanks to Jessica Dubin[3] for feedback on a draft of this post!

---

[3]https://www.linkedin.com/in/jessicalynndubin

3 I'm taking a sabbatical and attending Recurse Center!

Originally published on 2022.09.11.

It's been almost a decade since I graduated from college. In that time, I've worked at three startups, co-founded a non-profit immigration tech company, consulted for the United Nations, and noped out of grad school after one semester (twice!). I've also struggled with depression and anxiety, had three different therapists, and tried multiple different anxiety and depression medications. And I've adopted three cats, met and married my wife, and had two kids with her.

During that decade, I've kept learning. On the job. On the weekends. In my evenings. I'm tired.

During that decade, I've not had time to sit down and really dive deep into becoming a better programmer, a better software engineer. I've done a lot I'm proud of, but I haven't had the chance to dive deep since college. It's time to do that. I'm going to take a sabbatical from work to spend dedicated time becoming a better programmer and software engineer.

This is a great privilege, and not one I'm taking lightly. Many people do not have this opportunity for myriad reasons, and I'm grateful.

I'll be taking 12 weeks off of work and attending Recurse Center[1] in the Fall 2 batch, starting September 19th. Here's what I'll be learning and how I'll be doing it.

3.1 My Recurse Center Plan

My overall goals for attending Recurse Center are:

- Learn systems programming better
- Learn how things like key-value stores, databases, and queues work under the hood and what makes them efficient/performant
- Learn more effective debugging
- Learn how to performance profile things other than CPU

---

[1] https://www.recurse.com/

To specifically achieve these, I have a couple of project ideas that are in loose stages. I don't want to get too detailed in my planning lest I lose flexibility (a tip from multiple RC alumni!), but I need *some* plan or I'll spin my wheels. So, I'm going to work on two main pieces of software while I'm at RC:

- Key-value store compatible with (a subset of) Redis
- Chess engine

That second one is going to be rationalized as a way to understand performance optimization, low level stuff in general, and it'll also have some disk or other IO, but honestly... I also have just had a yearning to do it for so long. So I'll rationalize it, but let's be honest about why I really want to do it.

There is also a lot of pair programming as part of RC. I'm looking forward to learning from everyone else in the batch, and helping them in their learning journey however I can. Learning together is a tremendous way to make faster progress than learning alone. You also learn things you wouldn't have learned on your own. Serendipity is a tremendous thing.

If you want to follow along, everything will be open source. This is a requirement of RC so that people can collaborate, and I'm looking forward to learning in the open—but I'm also a bit nervous! Here are the repos I'll be working out of:

- Key-value store: https://github.com/ntietz/anode-kv
- Chess engine: https://github.com/ntietz/patzer

You can also follow me on GitHub[2] in general to see all the things I'm working on.

3.2 Why Recurse Center?

I am taking the time off work, but why attend Recurse Center specifically?

To benefit from the community, and to benefit the community. Going through this learning process with a group of peers who are also learning will help me stay on track and get unstuck when I inevitably run into barriers. And I'll learn unexpected things by helping other people, too! I've long wanted to attend after hearing about the experiences of folks like Julia Evans.

Now's the time, since it's still online (going to NYC for a bit would be disruptive for family life). I can't wait to pair program with a bunch of

---

[2]https://github.com/ntietz

great folks on their work and mine. And I hope to come out of it with some new friends.

4 RC Week 1: Getting Unexpected Extrovert Energy

Originally published on 2022.09.24.

The first week of my batch at Recurse Center[1] (RC) just finished, and it was an intense week! I'm planning to write a blog post each week about my experience at RC. They'll vary, but it'll probably be a mixture of what I did and my feelings about everything. There won't be *too* much technical content—I'm planning to write individual blog posts on specific things I'm learning.

If you're wondering why I'm attending, take a look at my blog post announcing my sabbatical[2].

4.1 How was the week?

The week was really great, and also a lot.

I am normally very drained by socializing, and this week had a ton of that. I was expecting to be very drained, because I identify as an introvert, and yet... I got this weird extrovert energy from all my conversations! I'd come out of each one charged up and ready to keep going and doing more and more! When I talked about this, I found out this is an experience shared by some others.

For me, I think it's the culture, the people, and just the sheer excitement of being here, but I'm not really sure at all. I want to learn two things: How can I capture this energy outside of RC at a day job? And how do I create this sort of culture in other places? I'm looking forward to seeing if this continues, wears off, or changes.

Here's what I did this week:

- Had 7+ coffee chats
- Went to 2 mixer meet and greets
- Had 4 pair programming sessions
- Attended a ton of events

---

[1] https://www.recurse.com/
[2] https://ntietz.com/blog/going-to-recurse-center/

- Formed a home lab discussion group
- Formed a Red Book[3] reading group
- Stood up a process for my KV store[4] (it does hardly anything, but it parses requests and responds with errors!)
- Started on my chess engine[5] and detoured into GUI programming to make something visual
- Implemented a couple of initial dummy chess strategies (first-legal-move and random-move)
- Started using Obsidian to take notes, and enjoyed it a lot
- Ordered a new-to-me used server (technically, a workstation) for use in my performance testing
- Figured out the source of my arm pain and resolved it. **I can use a keyboard full time again!**
- Got a cold and used a ton of tissues

My week was very weighted toward social things. I wanted to meet as many people as possible and try out all the events before focusing more. Next week I'm going to pare it down a little and dig deeper into my projects.

4.2 Takeaways

I learned a few things this week.

Pair programming is hard. It's really freaking hard. If you're not careful (and, reader, I'm not careful) it can turn into performance, and that is not what it's supposed to be! I'm working through some feelings of needing to be right, needing to not flail around or be ignorant when I'm the driver. Fellow Recursers have given me some really great advice on how to work on this. Next week, pairing will be a focus, and I am looking forward to the practice!

Recurse Center is a magical place. Yeah, I know, everyone says this. It's true, though. This is an amazingly supportive environment and I cannot imagine a better place to learn. But check back in with me in 11 more weeks. If there's a honeymoon phase, I'm still in it!

Plans are helpful, as long as you're flexible. This was advice given to me by some RC alumni: Come in with a plan, but be willing to deviate as you discover what you're interested in. I am *so* glad that people told me that, because I do have a tendency to stick to a plan and I can get anxious if I don't follow through. This week, the main way I bent my plan was by leaning into GUI programming. I've never made a native GUI

---

[3] http://www.redbook.io
[4] https://github.com/ntietz/anode-kv
[5] https://github.com/ntietz/patzer

before, and it was a very different and *very* fun experience. If someone hadn't told me explicitly to be flexible in my plans, I may not have done that!

GUI programming is really fun! I'm using egui[6], an immediate mode GUI library for Rust. It has been more intuitive for me than Qt or GTK were when I tried those, but that was also... over a decade ago. (Yikes.) It's definitely more intuitive for me than React. We'll see if it remains that way when state gets more complicated in my application!

4.3 What's next week?

Next week I'm going to cut back on the social side to make more time for pair programming. I'm going to go to the groups that are most relevant to what I want to learn at RC, and skip those that aren't as relevant.

I want to have a coffee chat with someone each day, and I want to pair program each day. These are "best effort" attempts. If I don't hit daily I won't feel bad, but that's the goal.

For my KV store, my goals are:

- Implement the ECHO, GET, and SET Redis commands
- Run a benchmark of those commands and compare to Redis
- Start learning how to figure out why they're slower than Redis! (Just going to assume they will be haha.)

For my chess engine, my goals are:

- Make the GUI interactive, so we can play the engine. (I think this will be easier than implementing UCI[7], but I'll do that eventually as well.)
- Implement a slightly-better search algorithm (requires implementing evaluation as well), like minimax

Along with those, I have a few ancillary things I'm going to work on:

- Install Proxmox[8] on my server and set up my home lab so I can drag race databases. I'm probably going to learn Ansible to manage it.
- Write a blog post on the paper I'm reading for the Red Book reading group
- Finish up a blog post draft I have sitting in my backlog, and maybe give a presentation on it at RC (it's about coding by voice!)

---

[6]https://github.com/emilk/egui
[7]https://www.chessprogramming.org/UCI
[8]https://en.wikipedia.org/wiki/Proxmox_Virtual_Environment

I think that's all I wanted to say about this week! If you read this and you're curious about RC, or you want to say hi, my email is down below.

5 Rounding in Python

Originally published on 2022.09.28.

In software engineering, there are two principles that often come into conflict. The first one is the principal of least surprise. The second one is doing the right thing. These come into conflict when the usual thing that people do is in fact the wrong thing. A particular example of this is the behavior of rounding.

In school we were taught that rounding is always done in one particular way. When you round a number it goes toward the nearest hole number, but if it ends in 5, than it goes toward the higher one. For example, 1.3 rounds to 1, and 1.7 rounds to 2. And we were taught that 1.5 rounds to 2, and 2.5 goes to 3.

Because this is the way that we were taught rounding works, it can be quite surprising when rounding works differently. In fact, there are a number of different ways to round numbers. The Wikipedia article on rounding[1] gives no fewer than 14 different methods of rounding. Fortunately, with computers, we expect fewer: The IEEE 754 standard for floating point numbers defines five rounding rules.

Those five rules, along with their Python equivalents, are:

- round toward infinity (`math.ceil`)
- round toward negative infinity (`math.floor`)
- round toward zero (`math.trunc`)
- round half-to-even (`round`)
- round half-away-from-0 (no built-in equivalent that I found)

Sneaking in there is `round`, defined as rounding half-to-even. A lot of people are surprised by this the first time they call `round` with Python! It definitely is surprising if you are expecting the "round half toward higher numbers" behavior.

```
>>> round(1.5)
2
>>> round(2.5)
2
```

---

[1] https://en.wikipedia.org/wiki/Rounding

So that we can see that Python's rounding behavior the principal least surprise. Why is this the default behavior?

There really two good reasons have rounding half-to-even as the default:

1. It's more likely what you actually want. When you always round up, you introduce bias across a lot of rounding operations. When you sum up the rounded values, you'll have a little bit less bias in the final sum.

   In fact, some of the Python docs[2] mention that floating point math guarantees rely on the half-even rounding in some cases:

   > The algorithm's accuracy depends on IEEE-754 arithmetic guarantees and the typical case where the rounding mode is half-even.

2. Having it as the default is... the standard. The IEEE 754 standard for floating point numbers requires this as the default.

   From the standard:

   > The roundTiesToEven rounding-direction attribute shall be the default rounding-direction attribute for results in binary formats.

Of course, the standard also requires that five different rounding mechanisms are available to users. Python does make those available, but only on the `decimal` type. The other expected behavior can typically be implemented using `floor`, `ceil`, and `trunc`. Of course, that's extra work and room to get things wrong.

At the end of the day, if your application depends on specific rounding behavior than you should probably verify what behavior your libraries give you before you use them. And, of course, Python does give you the functionality you need in the decimal[3] package. To quote the docs:

> The decimal module provides support for fast correctly rounded decimal floating point arithmetic.

It gives you all the rounding modes you want, more exact representations, and less error introduced into arithmetic. When you care about the details a *lot* and your application depends on them, you can get the rounding you want! And when you don't care about it, but just want the thing that probably works, Python gives you a reasonable default.

---

[2] https://docs.python.org/3/library/math.html#math.fsum
[3] https://docs.python.org/3/library/decimal.html#module-decimal

Ultimately, I think that the Python and choice here the break ties toward even numbers is a sensible choice, made stronger by the presence of the decimal package. Managing these tradeoffs is difficult, and the Python developer who chose this rounding behavior made the right call. I, for one, would rather have people accidentally do the right thing and be surprised, rather than avoid surprise so that people can do the wrong thing.

Extra content time! I did some sleuthing to see where and when this behavior came from. This is all "as far as I can tell"—if there are errors, please let me know nicely.

When was the round function added to Python?

It was added in commit 9e51f9bec85[4] by Guido van Rossum himself. The intial implementation:

```
static object *
builtin_round(self, args)
    object *self;
    object *args;
{
    extern double floor PROTO((double));
    extern double ceil PROTO((double));
    double x;
    double f;
    int ndigits = 0;
    int sign = 1;
    int i;
    if (!getargs(args, "d", &x)) {
        err_clear();
        if (!getargs(args, "(di)", &x, &ndigits))
            return NULL;
    }
    f = 1.0;
    for (i = ndigits; --i >= 0; )
        f = f*10.0;
    for (i = ndigits; ++i <= 0; )
        f = f*0.1;
    if (x >= 0.0)
        return newfloatobject(floor(x*f + 0.5) / f);
    else
```

---

[4]https://github.com/python/cpython/commit/9e51f9bec85

```
        return newfloatobject(ceil(x*f - 0.5) / f);
}
```

It looks like it was initially rounding half-away-from-zero! And it's pretty easy to read.

This was changed in 2007 by Guido van Rossum, Alex Martelli, and Keir Mierle in commit 2fa33db12b8cb6ec1dd1b87df6911e311d98457b[5]. Here you can see the now-more-complex implementation:

```
static PyObject *
float_round(PyObject *v, PyObject *args)
{
#define UNDEF_NDIGITS (-0x7fffffff) /* Unlikely ndigits value */
    double x;
    double f;
    double flr, cil;
    double rounded;
    int i;
    int ndigits = UNDEF_NDIGITS;

    if (!PyArg_ParseTuple(args, "|i", &ndigits))
        return NULL;

    x = PyFloat_AsDouble(v);

    if (ndigits != UNDEF_NDIGITS) {
        f = 1.0;
        i = abs(ndigits);
        while (--i >= 0)
            f = f*10.0;
        if (ndigits < 0)
            x /= f;
        else
            x *= f;
    }

    flr = floor(x);
    cil = ceil(x);

    if (x-flr > 0.5)
        rounded = cil;
    else if (x-flr == 0.5)
        rounded = fmod(flr, 2) == 0 ? flr : cil;
```

---

[5] https://github.com/python/cpython/commit/2fa33db12b8cb6ec1dd1b87df6911e311d98457b

```
    else
        rounded = flr;

    if (ndigits != UNDEF_NDIGITS) {
        if (ndigits < 0)
            rounded *= f;
        else
            rounded /= f;
        return PyFloat_FromDouble(rounded);
    }

    return PyLong_FromDouble(rounded);
#undef UNDEF_NDIGITS
}
```

Notably, we can see from the tags on GitHub that this was present in Python 2.7 and in Python 3.0. So, this behavior has been around for quite a while. There was quite some discussion[6] about it in the Python bug tracker at the time.

Well, our little historical escapade is over! I still agree with the folks in that discussion that round half-to-even is the right behavior.

Later!

There's a companion post to this one over on my friend John's blog! You can read his post[7] for another take on Python's rounding behavior.

---

[6] https://bugs.python.org/issue32956
[7] https://thetmpfiles.com/2022/09/28/why-is-python-rounding-wrong/

6 RC Week 2: Pairing is Awesome

Originally published on 2022.09.30.

The second week of my batch at Recurse Center[1] (RC) is a wrap, and it already feels like it's going too quickly. My batch is twelve weeks long, so I'm 17% through. Only ten weeks left! This is a precious time, so I'm trying to make the most of it, but also trying to not increase the pressure on myself to make the most of it. This can get a bit recursive, which is, ah, in the name I guess!

6.1 How was the week?

Overall, this week was pretty good. I was sick for a lot of the week (the joys of a toddler just entering preschool for the first time), which put a damper on my plans from last week[2]. The biggest wins of the week were learning and being kind to myself.

This week I had planned to get a lot of programming done. I got some done, but learned more than I maybe expected to, especially about myself and about pair programming.

Here's what I did this week:

- Had 6 coffee chats ☕
- Had 6 pair programming sessions, a mixture of working on my projects and working on other folks' projects
- Did some pair blogging, which was a fun and productive experiment!
- Went to a few events, like the Red Book[3] reading group, a homelab group (a blog post is coming on my homelab soon ☺)
- Figured out why egui[4] wasn't registering clicks as I thought it would (if you nest widgets, you can only interact with the one of them, I think the outer one) which unblocks me for progress on this next week!

---

[1] https://www.recurse.com/
[2] https://ntietz.com/blog/rc-week-1-recap/
[3] http://redbook.io
[4] https://github.com/emilk/egui

- Implemented COMMAND, ECHO, GET, and SET in anode-kv[5] and tested the performance; it's okay and single-threaded throughput is 0.75x to 1.15x redis's, depending on the workload. This gives me a very good launching off point for measuring performance, profiling, and making data-driven improvements!
- Summarized a paper for Red Book[6] reading group and presented it
- Setup my new server and got some automation running to provision VMs, yay!
- Scratched a personal itch and wrote a small Rust program to do a very specific task (filter a calendar feed to remove some cancelled events that showed up as phantoms in my Fastmail calendar) and got it into "production". From creating the repo to using it was ~2 hours, which felt fantastic. (Also the repo is 60% Rust and 40% Dockerfile, which I think is hilarious.)

Yeah, so overall, I think that I had a very productive week! I came nowhere close to the goals I wanted to get done this week, and I was productive, which means I was unrealistic, but more on that in the next session.

The biggest thing this week was all the pair programming. During pairing sessions, I learned things about my tooling that I didn't know. I learned about Rust features and Rust libraries that I didn't know. And I learned about little things that take us on tangents that are so wholly unrelated to programming, but just fantastic.

6.2 Takeaways

This week continued the unexpected side of RC for me: self discovery.

Pair programming is *awesome*. Before RC, I've been skeptical of pair programming. I've also been very afraid of it, as someone who is easily drained by social interactions. RC has flipped this on its head for me and showed me the joy of pair programming. I won't even say "pair programming done well," because I certainly don't think I know how to pair well yet. I think it's joyous when everyone is approaching it with kindness and openness. It's not always roses, but it has been instrumental in me learning so much this week.

I set too high of expectations for myself. Looking back at my list of things I wanted to do this week, it was way too ambitious: I wanted to implement a few redis features, benchmark them, setup a server, make an interactive GUI, implement an AI algorithm, write a blog post, finish another blog post, and pair program a ton. Yeah, that was not realistic.

---

[5] https://github.com/ntietz/anode-kv
[6] http://redbook.io

But is it a problem? In this context, in this week, it was not. I was able to be kind to myself and understand that not only was it *too much*, I was also sick.

In general, it is a pattern. I have a tendency to be hard on myself and set very high expectations for myself. The problem isn't necessarily the expectations, but if I make myself feel bad for falling short. If I just have high expectations, that can be an effective motivational device. So this week it worked out. Next week, I hope it does, as well.

6.3 What's next week?

I set too high of expectations this week, and it worked out. So let's do that again and play with fire, I guess?

For events/social things, I want to make sure that I pair program every day and keep having coffee chats. I have a few events I'm going to related to relevant topics. And I'm going to explore a few new ones that I wasn't able to make the time for this week! (I want to keep meeting more people who I haven't interacted with very much so far in the batch, and keep exploring different things!)

On specific projects, I do want to circle back to my chess programming and make progress on both.

For anode-kv[7], I want to:

- Finish up my VM automation for running benchmarks
- Benchmark and profile anode-kv and compare to redis under the same workload
- Make one improvement to performance based on data
- Learn to use rust-gdb and use it to find and fix a bug (I assume I have a bug to fix)
- Implement INCR and maybe some list commands (this will require me to refactor some of the storage layer to not just deal with `Vec<u8>`, but to have some tracking of what type a value is)

For patzer[8], I want to:

- Get the GUI interactive for humans, so you can play against a bot
- Implement one search algorithm like minimax (which will also require a basic evaluation function)

---

[7] https://github.com/ntietz/anode-kv
[8] https://github.com/ntietz/patzer

This is... a lot. I think there's a chance I will complete it all, but a relatively low chance. I'm okay with that! By giving myself a menu of things to work on, I can do what captures my interest at any given moment.

Alright. Time to get some rest (or finish up another blog post). If you read this far, hi! Thank you! I appreciate you!

7 Paper Review: Architecture of a Database System

Originally published on 2022.10.01.

Last week, I read "Architecture of a Database System"[1] for a Red Book[2] reading group.

This is as massive paper: 119 pages. What surprised me is how approachable it is. I have relatively little background building database systems and more experience using them. Despite this, the paper was readable and I was able to take away quite a bit from it, which I've already put into practice in my redis-compatible KV store[3] that I'm building to learn about database systems.

The paper is structured in a way that makes it easy to skip around and focus on the parts that are most interesting or useful to you at the moment. It also gives a lot of pointers into other papers or texts to learn more or build a foundation.

- The first section is under ten pages and gives a map of the rest of the paper as well as of architecture in general, so you can put the different pieces in context. This is probably the section I would recommend *everyone* read.
- The second and third sections are also really useful as a user of a database system to put in concrete terms why, for example, PostgreSQL does not handle large numbers of open connections very well. (Hello, PgBouncer[4]!).
- The fourth section gives an overview of the relational query processor and helps understand how queries are parsed, optimized, and executed.
- The fifth section talks about storage and what considerations go into making it efficient.
- The sixth section talks about transactions, concurrency, and recovery. This section breaks down what ACID is (spoiler: it's not well defined, but it's useful anyway), talks about locking, and most importantly goes through transaction isolation levels. It wraps up with durability. This section was probably the most intense for me!

---

[1] https://scholar.google.com/scholar?cluster=11466590537214723805
[2] http://redbook.io
[3] https://github.com/ntietz/anode-kv
[4] https://www.pgbouncer.org/

- The seventh section talks about the junk drawer that exists in database architectures, just like in all architectures: shared components that get shoved into one category, the section of misfit toys. I skimmed this one.

I think this paper is an excellent introduction to database architecture for users of databases and for anyone who wants to learn more about the internals. It will give you a good, broad foundation which you can use to drive further exploration and improve your understanding of databases as you use them.

8 Starting my (overkill) homelab

Originally published on 2022.10.06.

I've set up a homelab finally! This is something I've wanted for a while and finally the timing was right. The right project came along to justify it, so I took the plunge.

Naturally, that leads to a few questions: What's a home lab? Why do you want one? And what is the shiny hardware? (That last one is the dessert if you get through the rest ☺.)

8.1 What's a homelab?

Oh, and is it "homelab" or "home lab"? Google Search Trends seem to indicate that it should be "homelab" in this context. When people search for "home lab" it's about science labs or dogs. When people search for "homelab" it's about computer stuff. So, that's the spelling I'll be using!

A homelab is, simply put, a place to experiment with technology at home! It can take a variety of forms, but that's the crux of it. Some people have homelabs of rackmount hardware, but it can be as simple as an old laptop or a Raspberry Pi.

Ultimately, people use them for a variety of different things. From "lab" in the word, you figure that it's a place for experimentation, not production. That doesn't always end up as the case: Some people run "home production" instead of "homelab", but it's a big tent. All are welcome.

There are basically two camps that I see homelabs fall into:

- Used for hosting services: You'll see a lot of people hosting things for their own use. Things like Pi-hole[1] for ad blocking) and photo sharing apps fall into this bucket. There's a lot of emphasis in this group on frugality and deal seeking.

- Used for experimentation: You'll see people doing infrastructure and DevOps, security experiments, etc. This is the camp I fall into. I don't want my homelab to run anything that someone depends on. I'm also willing to spend a little more on it for expediency.

---

[1] https://pi-hole.net/

8.2 Why do I want one?

I mean, servers are fun. But I did say the right project came along to justify it. Or to rationalize it, at least.

Right now I'm attending Recurse Center[2], and one of my projects is making a key-value store[3]. When I benchmark even just the protocol parser on my laptop, timing varies by 50% if Zoom is open. This makes for a bad test. I want to have a consistent test environment that I can run benchmarks on.

I would also like to have extra computing power to throw at things like tournaments of chess engines, since I'm working on a chess engine[4] at RC as well.

I could do both of these using cloud servers, but... that can get expensive pretty quickly. This way I have an always-available local box that I pay for once and can use (and abuse). As a comparison, an equivalent AWS instance to what I got would cost the same amount after being on for 20 days. I could turn it on only when I need it, but in the long run having this machine is much more cost effective.

Also the hardware is shiny, and it makes my Rust builds significantly faster (30 seconds instead of 50 seconds for a full rebuild).

8.3 Okay, so what's that shiny hardware?

Okay, okay, I'll tell you about the hardware. I got a used Dell T7910[5] from PC Server and Parts[6]. This was originally a video editing workstation, so it has a ton of PCI available, which will be handy for storage. Inside it, it has 2 Intel Xeon E5-2690 v4 CPUs (28 cores / 56 threads) and 128 GB DDR4 registered ECC RAM. I also have a paltry 1 TB SSD, which I plan on using as the boot drive but eventually augmenting with spinning disks and NVMe SSDs in those totally extra PCI slots.

"But Nicole, that's ridiculous. That hardware is so overkill."

Yes, you're not wrong.

But it's shiny, and it'll last me a long time. It's accelerating my Rust builds and that's actually a real productivity boost. And it was cheap.

---

[2] https://www.recurse.com/
[3] https://github.com/ntietz/anode-kv
[4] https://github.com/ntietz/patzer
[5] https://www.dell.com/en-us/shop/workstations-isv-certified/precision-7910/spd/precision-t7910-workstation
[6] https://pcserverandparts.com/

It's a used workstation from 2017, so it's much much cheaper than new hardware.

8.4 How did you set it up?

Okay, so this is the saga, and it was painful. At first, I wanted to set it up with Proxmox so that I can provision VMs. And I wanted to do that provisioning with Terraform and Ansible. This wasn't an entirely spurious decision: I wanted to be able to benchmark things in isolation, and thought that would be a good approach.

Ultimately, it was a very painful decision, though. I never *quite* got things working and I was having really weird network issues, where I could not initiate outgoing connections from the VMs. And eventually I broke things so badly, I couldn't even ssh into them!

In the end, I realized that I didn't... really... need to do this? Why am I doing Proxmox again? Oh right, because I thought it was the "right" way to do it. No, no, that's okay. I'll do things the quick and dirty way 😀.

Instead of all the fuss, I just installed Fedora on it and called it a day. If I want VMs, I can explore things like Firecracker in the future, too!

Well, that was the end of the first part. But I couldn't get the thing to boot consistently! I had to mash F12 to get into the boot menu every time, because the drives connect through a SAS RAID controller. This controller permits direct mount of drives which aren't in RAID, but it seemingly does *not* permit default booting off one. Sigh. This is a big problem for me! If the power goes of and I'm away, I want to be able to wake the computer with a WOL packet, then have it come back up.

Fortunately, the solution was pretty straightforward. A fellow Recurser pointed out I should probably just connect that drive directly with SATA instead of through the SAS array controller. I could... but there are only two SATA ports on my mainboard and they're both in use! It turned out that the second one was not in use, but had a cable plugged in and cable managed into the back of the case for exactly this sort of use case. I plugged in my boot drive to that instead and everything came on! Reboots are consistent now, and it always comes back on.

8.5 Should you make a homelab?

I don't know! It depends. If you have something you want to use it for, by all means! It can be rewarding and useful. It can also be a new hobby that sucks up a bunch of your time and money.

Gamble wisely.

Thanks to fellow Recurser Mikael Lindqvist who paired with me on writing this blog post! It was a super fun way to get a post written, and I'd recommend trying it.

9 RC Week 3: Returning to Math

Originally published on 2022.10.08.

The third week of my batch at Recurse Center[1] is finished. It is still flying by too quickly. Nine weeks left!

This week was a whirlwind and really busy. I think I pushed myself too hard. I had just recovered from my cold and was a little drained, and then got my COVID booster and flu shot, which really knocked me out. But I got a lot done, and I'm going to focus on self-care a little bit more next week.

The most exciting thing this week is probably that I'm taking a turn back toward math! I joined the category theory group at RC ("Category Theory Catacombs") where we're working through the Programming with Categories[2] course. This was a little intimidating to me before I started it, because it's been the better part of a decade since I have attended a math lecture. For some reason I decided to join since I was taking an easy day "off" after my vaccine, so I would just watch some math lectures.

It turned out to be the highlight of my week and reminded me the joy of approaching things from a mathematical perspective. I need more of this in my life, so a few of us are also going to start exploring theorem provers[3] and working through Theorem Proving in Lean[4]. I'm so excited to continue going down this path.

Besides that, I did get some things done that I'm pretty proud of:

- Implemented INCR/DECR in anode-kv[5]
- Benchmarked anode-kv against redis and found very favorable results (we will see how these hold up over time though, as features get added!)
- Had at least one pairing session a day and at least one coffee chat a day
- Wrapped up the first pass[6] at a GUI frontend for patzer[7] (my chess program)

---

[1] https://www.recurse.com/
[2] http://brendanfong.com/programmingcats.html
[3] https://en.wikipedia.org/wiki/Proof_assistant
[4] https://leanprover.github.io/theorem_proving_in_lean/
[5] https://github.com/ntietz/anode-kv
[6] https://github.com/ntietz/patzer/pull/1
[7] https://github.com/ntietz/patzer

- Fixed an irritating boot drive issue with my server, so now it doesn't need to be babysat after a reboot

I have some capital-T Thoughts about immediate mode GUIs now, and the particular one I'm using, but those are best saved for another blog post that may or may not come to fruition. I don't know how coherent those thoughts are, but I'm pretty frustrated right now with this library. I'm proud of where I've gotten with it, though!

I'm also really proud of anode-kv so far. The core architecture is based on what I read in Architecture of a Database System[8], and it seems to be effective! Right now it can pass 1.7 GB/s through it in my test environment, contrasting with 360 MB/s for redis (with durability off, for a more fair comparison). The bulk of the time is spent in network syscalls and memory allocation/deallocation. I think there's room to speed things up, but also... there will be more pressing, more important performance problems after implementing durable storage[9].

One of my big takeaways with my performance work at RC so far has been reinforcing a couple of things I've heard before: - **Estimate bounds before you benchmark:** Estimating theoretical bounds or pragmatic "good enough" bounds before benchmarking is helpful for understanding both (1) if your benchmark is working, and (2) if it indicates good enough or a problem. - **Profile before you optimize:** I expected the memory copies that I do to be expensive, but it turns out that they're not so bad in terms of the overall performance.

It might be a fun exercise to profile redis itself to see what it's doing that's making it slower than anode-kv. Maybe that's on the docket for the next week or two!

9.1 What's next week?

Okay, so next week I want to:

- Keep pairing every day, keep coffee chats every day
- Make progress on patzer[10]:
  - Implement basic board evaluation (dumb strategy first: material count)
  - Implement one slightly better search algorithm like minimax
- Make progress on anode-kv[11]:

---

[8]https://ntietz.com/blog/review-architecture-of-a-database-system/
[9]https://github.com/ntietz/anode-kv/pull/5
[10]https://github.com/ntietz/patzer
[11]https://github.com/ntietz/anode-kv

- Implement durable storage and see how it performs (naive implementation first!)
- Implement set and hash operations

- Read another Red Book[12] paper
- Learn some category theory
- Learn some Lean
- Go to some of the fun/weird/quirky programming events

But I'm going to be more flexible than usual with my plans next week. I'll be visiting family in Ohio, and my wife's going to a conference, so I might be off kilter or interrupted more. And that's fine!

That's all for this week. It's been a long week, and I have to go pack up to travel tomorrow and get some sleep. I hope you have a great rest of your weekend!

---

[12] http://redbook.io

10 Paper review: The Gamma Database Project

Originally published on 2022.10.11.

Last week, I read "The Gamma Database Project"[1] for a Red Book[2] reading group. Unlike the last paper[3] for this group, this one was a lot more approachable in length: 19 pages.

I'm putting up some of my notes here from reading the paper. If you read through to the end, there's dessert: a quibble I have with the paper.

My understanding is that this paper was very influential in its time. The architecture it describes is a shared-nothing architecture for distributed databases with very nice scaling properties. Notably, it has linear scale-up and speed-up. These are often related, but they're distinct and both are important to examine.

- **Speed up** here is measuring how much faster particular queries get if we add more hardware. Since Gamma shows linear speed up it means that if we go from 5 to 10 machines, we should see queries run in half the time.
- **Scale up** here is measuring how much data can be handled by the system with fixed query times. Since Gamma shows linear scale up, it means that if we double the amount of data stored, and we double the machines in the cluster, then we should keep the same speed of queries.

They're related, but not the same: If a query is only hitting one server, adding more servers won't speed it up, for example.

They presented three key techniques for achieving these properties on commodity hardware:

- Horizontally partitioning data across the cluster (with some nice resiliency properties)
- A good parallel join function based on hashing

---

[1] https://scholar.google.com/scholar?cluster=8912521541627865753
[2] http://redbook.io
[3] https://ntietz.com/blog/review-architecture-of-a-database-system/

- Effective scheduling of jobs onto machines to make use of all available hardware

---

The overall architecture is pretty typical for databases; we can refer back to the Architecture of a Database System[4] for the overall architecture and just talk about differences.

The main differences come down to partitioning. They employ three different partitioning schemes:

- Round-robin: this is the default, and distributes records uniformly across all disk drives. This means that any read *must hit all disk drives*.

- Hashed: a hash function is applied to the input to determine which node gets the data.

- Range partitioned: the operator may select which range of data goes to which machines.

Hashed or range partitioned data may hit a subset of machines for queries, which has benefits (potentially more parallel queries, could be faster, less overhead from distribution) and has some drawbacks (more limitation in speedup and scaleup).

They do say that defaulting to splitting all data across all machines by default was a mistake, and that it would be better to base the amount of distribution on some metric. I wasn't clear on *why* they felt it was a mistake, so I'd like to learn more there.

I went on a nice rabbit hole exploration during reading this paper.

They mentioned they were using commodity hardware, so I was curious what the hardware was and how it has changed to today. It used cutting edge hardware at the time (they complain about being beta testers for some of it, delaying their project), and today it can largely be beat by a single desktop computer. My workstation has nearly as much *CPU cache* as the cluster had main memory.

The paper was released in 1990 but the hardware was acquired in 1988, so that's 34 years ago. Hardware today should be about 2^17 times "better", or 131,000x, but this may be on multiple axes (both increases

---

[4]https://ntietz.com/blog/review-architecture-of-a-database-system

in performance and decreases in cost, etc.). (Yes, I know Moore's Law has ended. Don't @ me.)

The hardware they had was 30x Intel 80386 processors, which ran at 16 MHz (one core). (Incidentally, these were still manufactured until 2007[5], as they were used in embedded applications long after personal computers outgrew them.)

Unfortunately, I can't find much information on cost of this system, but a simliar system was about $300,000 (about $700,000 in 2022). I can buy a system with at least 100x the processing power for at least 1/1000 of the cost, which would be 100,000x improvement, which is right about on the mark for Moore's law!

Saving the beef for last. They had one comment that seemed like mostly an aside, but I feel is not well supported. They state:

> "[...] the response time for each of the five selection queries remains almost constant. The slight increase in response time is due to the overhead of initiating a selection and store operation at each site."

I have a few issues with this claim:

- They don't provide a word on how much overhead these operations are (and I'm skeptical that they're high enough overhead to see this effect)
- The increases are not consistent! The times go up, then down, then up again, and it varies with respect to the query being run.

It's not even clear that the experimenters ran the queries multiple times and averaged the results. There's little information on how they gathered this data.

I think there's a much simpler explanation of this relatively minor variance: Simple probability.

In this case, they're increasing the number of nodes from 5 to 30. The operations require data from all nodes to return before they can be finalized. This means that the operation will be as slow as the *slowest node*. If you take the maximum of 5 random numbers in a range, and then you take the maximum of 30 random numbers in a range, you would generally expect the latter to be higher than the former—but not always!

At any rate, this doesn't really take away from what's an excellent paper to read.

---

[5] https://handwiki.org/wiki/Engineering:Intel_80386

11 RC Week 4: Gratitude and emotions

Originally published on 2022.10.14.

Wow, my RC batch is one-third done. I've just finished my fourth week, and there are eight weeks left. Time is flying by. I feel like I've settled into a decent groove.

Taking a step back, it is setting in how much I've learned so far and how much I've accomplished. In these four weeks, I've learned about the architecture of databases and managed to write a key-value store that has durable storage[1] and still outperforms redis on my machines. (It's multithreaded against redis's single thread, but that's their design choice.) I've also learned about how chess engines work and wrote one that, using a standard technique[2], can beat me.

The most important things, though, aren't these accomplishments directly. They're the feelings I've gotten.

Before my batch, I wasn't really confident that I could "do" database stuff. Sure, I can write stuff at work that performs well, but can I really do this deep hard tech? That's what **real** engineers do, not me! And can I write a chess engine? Gee, that's what **real**, hardcore engineers do. I'm not that hardcore!

Before my batch, I was also feeling pretty... I hesitate to say burnt out, but I was finding absolutely no joy in using computers. I would find things I wanted to read, wanted to learn about, and my brain would not kick into gear, would not engage. Programming hurt, and computers hurt. (And not just physically from my painful nerve/inflammation issue!)

But now, I'm feeling a lot more confident in all of this. First and foremost: **I fucking love coding again** and oh god, it's so much fun to write code. (I could qualify that, but no, it's just fun to write code!) And I'm also a lot more confident in my ability to *learn* now. I've read a few database papers, including a survey paper of over 100 pages, and got real tangible insights out of them. In a few weeks, I've gone from not being sure how something like redis works, to being able to (roughly) describe the architecture of databases in general. If I want to work on databases, and I'm engaged with it, I can do it.

---

[1] https://github.com/ntietz/anode-kv/pull/5
[2] https://github.com/ntietz/patzer/pull/3

11.1 The gratitude part

One of my fellow Recursers posted today that they feel lucky to have the time and space to explore things here, to have support from people, to be able to support others. I read this and realized that I haven't expressed this gratitude recently.

I'm really fortunate to have a great employer[3] who graciously let me take a sabbatical to go do this wild thing that I couldn't always even really explain.

I'm so fortunate that this program even exists. The faculty here are so dedicated to keeping the culture one that is warm, welcoming, supportive, where we can fully engage and where we can be vulnerable and learn together. The fellow Recursers here are so generous with their learning and with their time. I'm very fortunate to have landed in this place with so many other people.

I think I have some new friends, and hopefully friendships that will last for a while. The people here are fantastic. (The fact that they appreciate my puns doesn't hurt, either.)

One year ago, I was in an extremely rough mental spot. When I went for a walk this evening, I was struck by just how different my mental state is now than one year ago. I had recovered from that episode before RC, but RC has elevated me to the other end of the spectrum. When I was on a coffee date with a friend this morning, he commented on how much energy I have.

I'm very grateful to have been welcomed and accepted into this community where I can blossom as a programmer and as a human.

And I'm grateful to my family, who have been immensely understanding and supportive of this adventure, and have provided immeasurable help with childcare. Thank you, Eugenia, mom, and dad. And thank you, Sophia and Alexei, for understanding that mom is at work a lot.

11.2 I'm not crying, you're crying

Emotions. Whoops.

---

[3]https://www.remesh.ai/

11.3 Okay, what's next week?

Well, this week I got a few major things done (durability in anode-kv and alpha-beta negamax in patzer). Next week I'm going to set the stage for the next round of major progress, but it'll be smaller features and cleanup.

- Keep pairing every day, keep coffee chats every day
- Make progress on patzer[4]:
    - Write a blog post explaining alpha-beta pruning, mostly so that I can shore up my understanding of it!
    - Implement the UCI protocol (#5[5]) so I can start evaluating patzer against other engines
- Make progress on anode-kv[6]:
    - Implement set operations (#10[7])
    - Add tracing (#8[8])
    - Add command-line options and config (#7[9])
    - Clean up the transaction log handling (#6[10])
    - Have some preliminary discussions around how I would implement something like a relational DB on top of this foundation
- Read another Red Book[11] paper
- Read two chess papers (one on Deep Blue, one on Alpha Zero)
- Engage my math brain
    - Learn some category theory
    - Learn some Lean (chapter 3, so we're getting into proofs now!!!)

There's a lot next week! It'll be fun, and I fully expect that like most weeks, I've planned more than I can do. That's worked out so far, because it always gives me something to latch onto if the current thing is getting hard to focus on.

That's all for this week!

---

[4] https://github.com/ntietz/patzer
[5] https://github.com/ntietz/patzer/issues/5
[6] https://github.com/ntietz/anode-kv
[7] https://github.com/ntietz/anode-kv/issues/10
[8] https://github.com/ntietz/anode-kv/issues/8
[9] https://github.com/ntietz/anode-kv/issues/7
[10] https://github.com/ntietz/anode-kv/issues/6
[11] http://redbook.io

12 Alpha-beta pruning illustrated by the smothered mate

Originally published on 2022.10.18.

I've been working on Patzer[1], a chess engine, during my time at RC. The first engine-like thing I implemented for it was alpha-beta pruning[2], which is a way of pruning out branches of the search tree to significantly speed up search. This is a common algorithm, which I also implemented in my undergrad AI class. That doesn't mean that I fully understood it as I wrote it! It's pretty tricky in the details and not immediately obvious *why* the pruning works.

In the process of writing it and debugging it, another Recurser and I traced through the execution with a known position where we could calculate the execution. This let us figure out what was going wrong, and also gain some intuition for what the algorithm was doing. I'm going to use that same position here to illustrated alpha-beta pruning. (This is partially so that when I inevitably forget the details, I can come back here and refresh myself!)

We'll start with an overview of the algorithm viewed through the lens of the algorithm it enhances, minimax. Then we will look at the alpha-beta pruning algorithm itself. We'll wrap up by looking at our example position, a hand-constructed position which utilizes a smothered mate, and see how minimax and alpha-beta pruning work on it.

12.1 Minimax algorithm

The base algorithm we're using here is called Minimax[3], and the name comes from what you're trying to do: You want to *minimize* the cost of the worst case *maximum* cost.

The intuition here is that under best play, if your opponent is making optimal moves, then they're going to make the moves which put you in the worst possible position. You're trying to pick moves which make your worst case less bad. (And that's ultimately what playing good chess is

---

[1] https://github.com/ntietz/patzer
[2] https://www.chessprogramming.org/Alpha-Beta
[3] https://en.wikipedia.org/wiki/Minimax

about: not making mistakes, and taking advantage of your opponent's mistakes.)

Here's Python-esque pseudocode which we could use for a basic Minimax implementation:

```
def max_step(board, depth):
  if depth == 0:
    return score(board)

  max_score = INT_MIN;

  for move in board.moves():
    next_position = board.make_move(move)
    score = min_step(next_position, depth - 1)
    if score > max_score:
      max_score = score

  return max_score


def min_step(board, depth):
  if depth == 0:
    return -1 * score(board)

  min_score = INT_MAX

  for move in board.moves():
    next_position = board.make_move(move)
    score = max_step(next_position, depth - 1)
    if score < min_score:
      min_score = score

  return min_score
```

The `max_step` function is the one that corresponds to the current player: They're trying to maximize among their possible outcomes. The `min_-step` function corresponds to the opponent, who is trying to minimize their opponent's best case.

(As an aside: this is usually written in the Negamax[4] style, which reduces it down to one function. This is how I've implemented it in Patzer, but for clarity I'm presenting it as two separate functions.)

This algorithm will find all the best moves! Unfortunately, it's also slow. It exhaustively explores the entire state space of the game tree. For chess,

---

[4]https://www.chessprogramming.org/Negamax

this gets quite large quite quickly: There are over 6 trillion leaves in the minimax tree after the first 4 complete moves of the game (depth 8). My computer would not ever reach this depth.

So, what are we to do?

12.2 Alpha-beta pruning

This is where alpha-beta pruning comes in. It's an optimization for minimax which allows us to prune out major swaths of the search tree.

The core idea of alpha-beta pruning is that there are some branches we know we won't explore, because they're too good or too bad. If a branch has a way that we can guarantee a better outcome than another branch, our opponent won't let us pursue that. If a branch has a way that our opponent can guarantee us a worse outcome than another branch, we won't go down that one, either.

To make this work, we keep track of the lower-bound (alpha) and upper-bound (beta), which let us then eliminate branches once we've confirmed that the branch will violate one of the bounds that we can otherwise guarantee. Note that this is done depth-first, like minimax. This is crucial for finding a leaf quickly to evaluate.

Here's the pseudocode of the algorithm. Again, this is the two-function implementation; you can make it one function at the expense of some readability. I've put some inline comments to highlight the differences between this and minimax. These comments are only in the max step function, but apply equally to both.

```
# we add two parameters, alpha and beta, which track lower and upper
↪  bounds
def alphabeta_max_step(alpha, beta, board, depth):
  if depth == 0:
    return score(board)

  # note that we're not tracking the max or min anymore!
  # these are tracked via alpha and beta now.

  for move in board.moves():
    next_position = board.make_move(move)
    score = alphabeta_min_step(alpha, beta, next_position, depth -
↪  1)

    # when the score is higher than the upper bound, we just fail to
↪  the
```

```
  # already established upper bound.
  if score >= beta:
    return beta

  # when we find a score that's higher than alpha, our lower
  ↪  bound, we
  # can adopt it as the new lower bound since we know we can
  ↪  achieve
  if score > alpha:
    alpha = score

return alpha


def alphabeta_min_step(alpha, beta, board, depth):
  if depth == 0:
    return -1 * score(board)

  for move in board.moves():
    next_position = board.make_move(move)
    score = alphabeta_max_step(alpha, beta, next_position, depth -
↪  1)

    if score <= alpha:
      return alpha

    if score < beta:
      beta = score

  return min_score
```

Using these bounds turns out to be very helpful. Analysis on Chess Programming Wiki[5] indicates that this can cut down the search tree significantly. If we always get the best move first, we would only have to evaluate 5 million positions. Obviously we can't know what the best move is or we would just play it! But there are ways we can order moves to find the best move earlier, and if you order them randomly you will still prune significantly.

---

[5] https://www.chessprogramming.org/Alpha-Beta

12.3 Alpha-Beta Pruning Illustrated

Alpha-beta pruning is pretty dense and hard for me to understand without tracing through a position in a game tree, so let's use an example to do that.

Here is our starting position, with white to move.

This position is set up for a classic checkmate known as the smothered mate[6]. From here, it's forced mate in two if white finds the right moves. We chose this position as our starting position since there is a clear tactical solution which is easy to evaluate as a human, and because it's a treacherous position: If you make the *wrong* move, black has checkmate available as well.

---

[6]https://en.wikipedia.org/wiki/Smothered_mate

We're going to look at a dramatically simplified game tree, too, to illustrate the algorithm. For this illustration, we are just considering three branches (moves are numbered starting from 1 for clarity):

1. **The smothered mate line.** 1. Qg8+ Rxg8 2. Nf7#
2. **The "whoops I lost" line.** 1. Qb7 Rc1#
3. **The pawn line.** 1. h3 h6 (Now no one has back rank mate) 2. h4 h5

We'll look at each of these first moves by white, followed by black's possible follow-ups, to a maximum depth of 4.

Here's our game tree, where we're examining at most 2 branches each time, except the root:

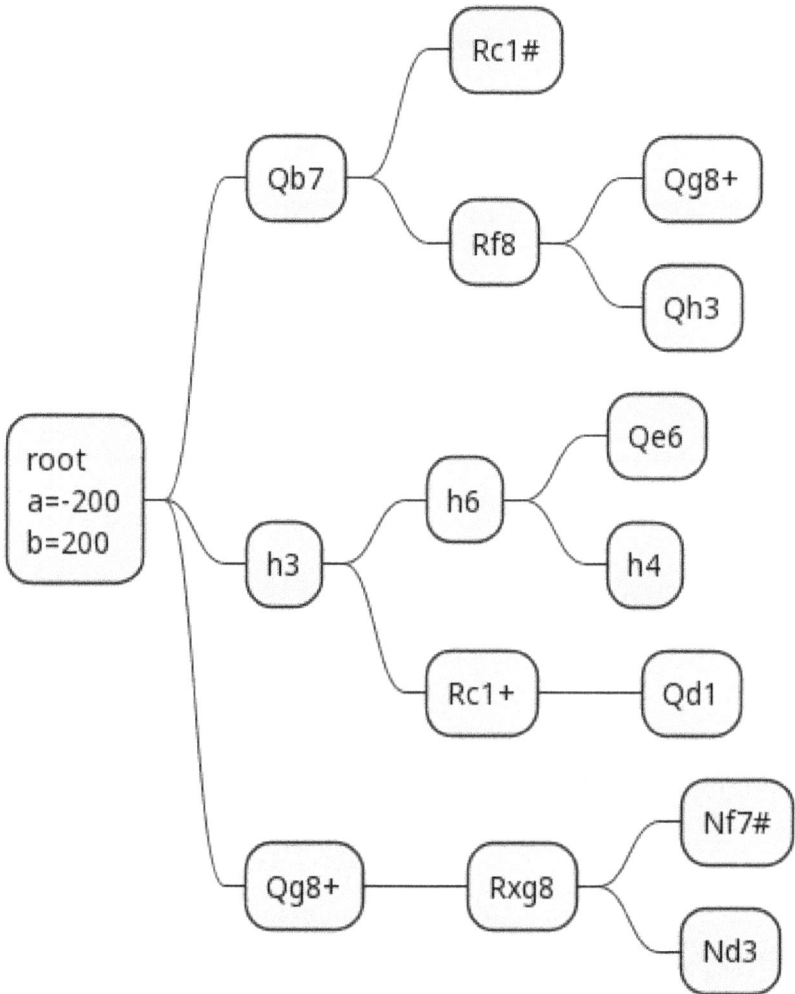

This is as dramatically reduced game tree, and it's already fairly overwhelming!

(If you're not used to the notation and you're wondering what things like "Qg8+" mean, this is algebraic notation[7]. It's not the most intuitive, but it's standard, so I think most chess players will be able to read it.)

If we look through this with minimax, we'll find one forced mate! There's

---

[7]https://en.wikipedia.org/wiki/Algebraic_notation_(chess)

another one hiding out, but it's not in our tree here, so we wouldn't find it. Since this move tree is for the player to move, the first and third layers are what we are looking at. If we find a checkmate there, great! The second layer is what our opponent is looking at, and if they find a checkmate there, boo, bad for us.

With minimax, we'd evaluate 9 leaf nodes.

Now let's consider what we'd be able to do with alpha-beta pruning. Assume we start by evaluating the Qb7 line. Then we see our opponent is able to make the move Rc1#, and that results in checkmate! This means we don't have to evaluate any further down that move tree.

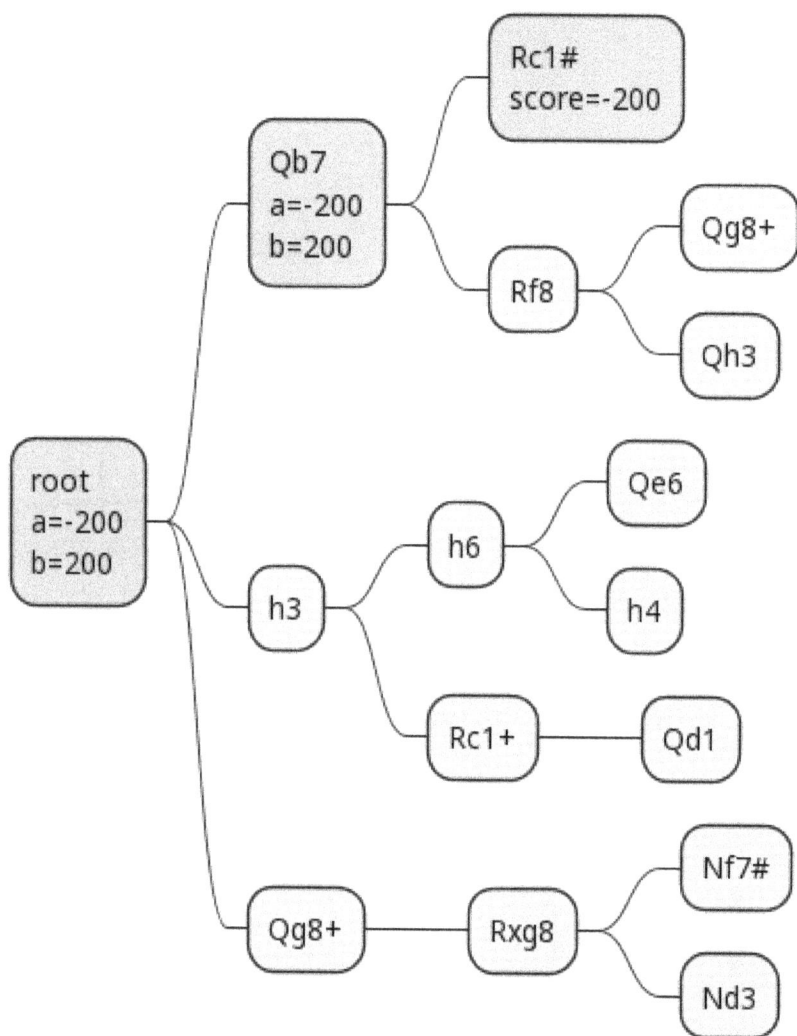

We score a checkmate for ourselves as 200 points, and a checkmate for our opponent as -200.

So after we evaluate the Rc1# position as -200, that hits our `score <= alpha` case for our opponent (`alphabeta_min_step`). This returns early, and our opponent prunes out the rest of that tree, since we know that there won't be anything better than a win for our opponent. We've gotten our first pruning!

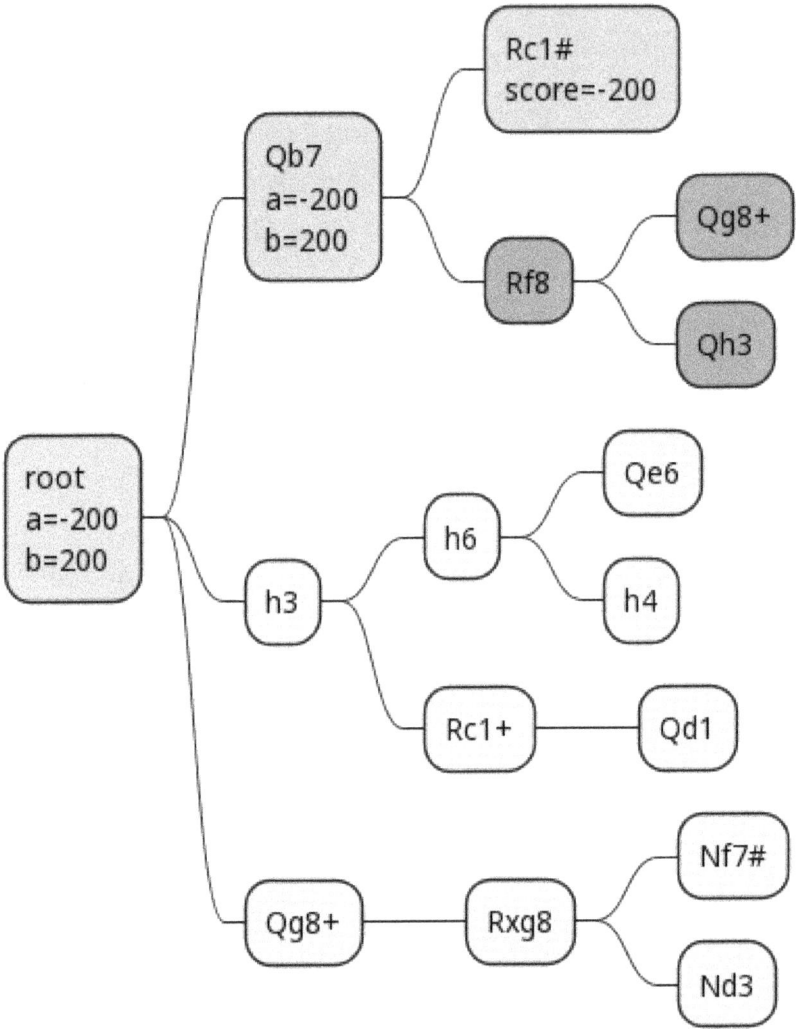

This has resulted in **no change** to our alpha and beta values.

Now we happen to pick h3 as our next move. We go down one line, and end up evaluating the final position as +10 (we have 17 points of material to our opponent's 7).

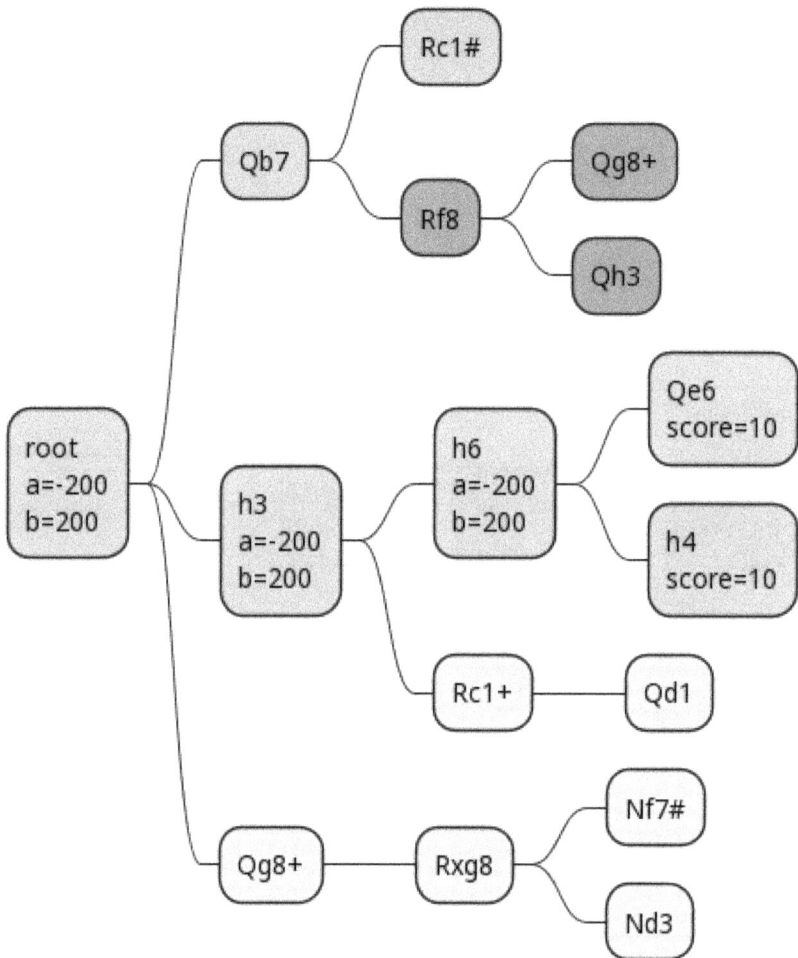

Unfortunately, this doesn't result in any pruning, but it does change the alpha and beta values.

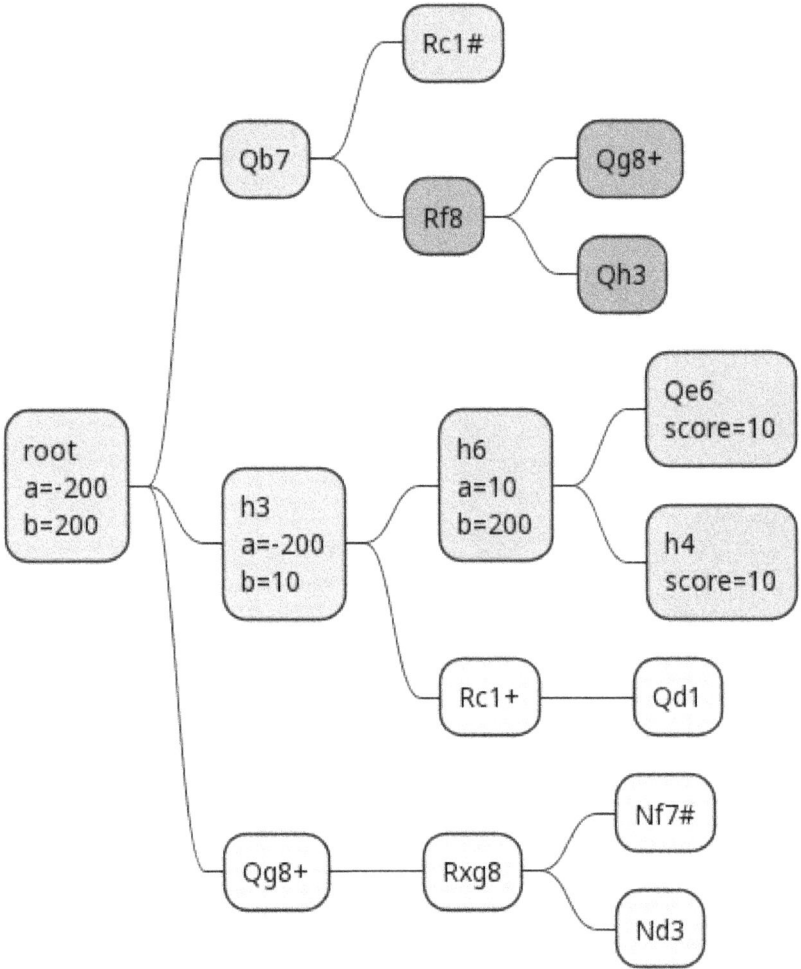

We finish out this branch and have a final score for it of 10.

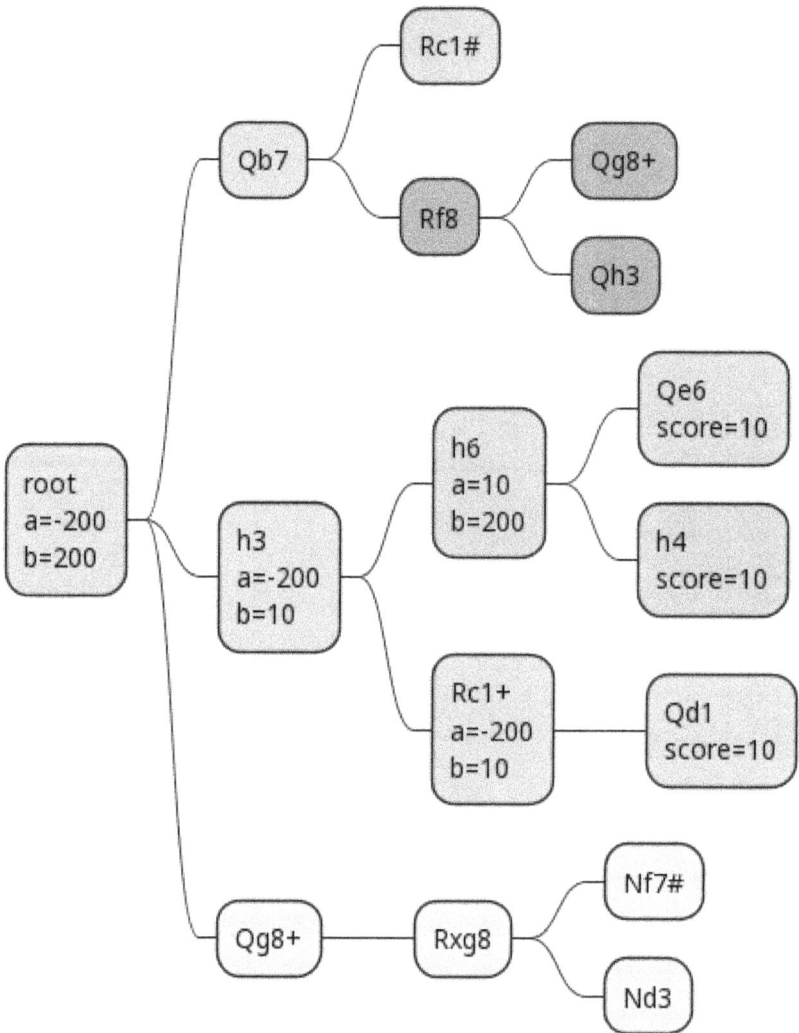

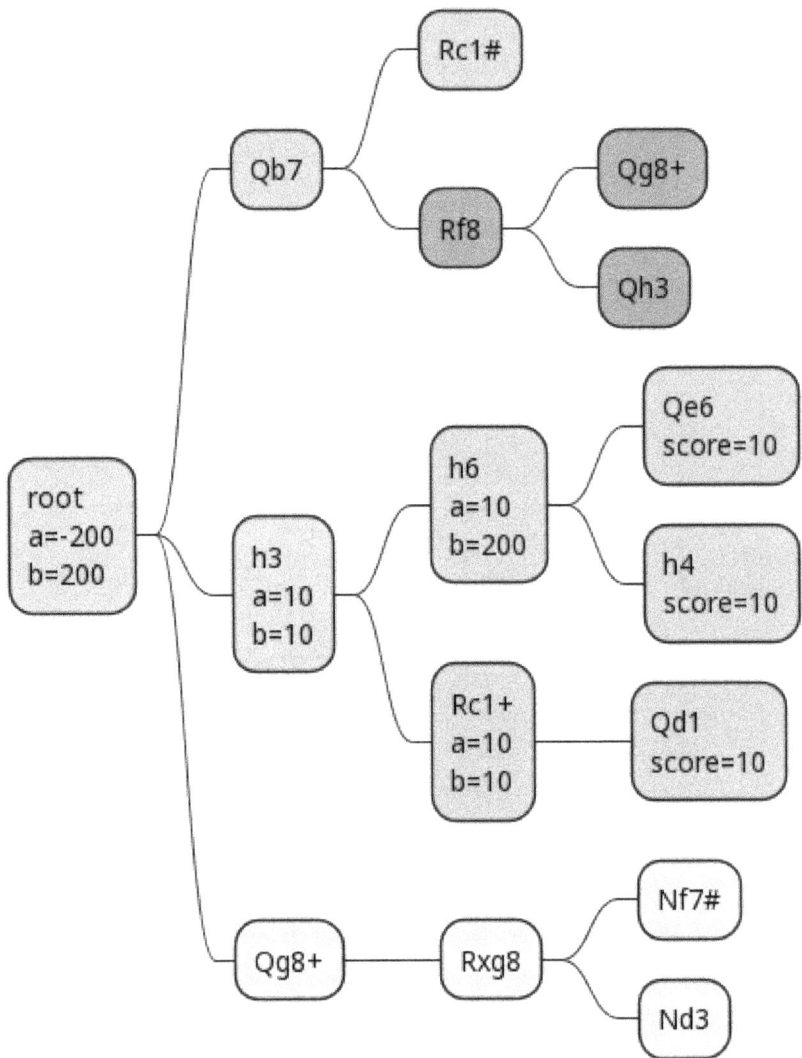

Here we can see that alpha and beta have both gone to 10, so even if we had a third branch to explore after h3, we'd know the score *must* be 10, because the upper and lower bounds have converged! This gives white the ability to guarantee at least 10 points, so alpha changes to that at the root. It's our lower bound.

Note: there is a line in there which at depth 4 does result in checkmate for our opponent. We didn't see it. This is one of the perils of evaluating to a particular depth, and there are techniques like quiescence search[8] which mitigate this. But we'll just pretend it isn't an issue here, and move on!

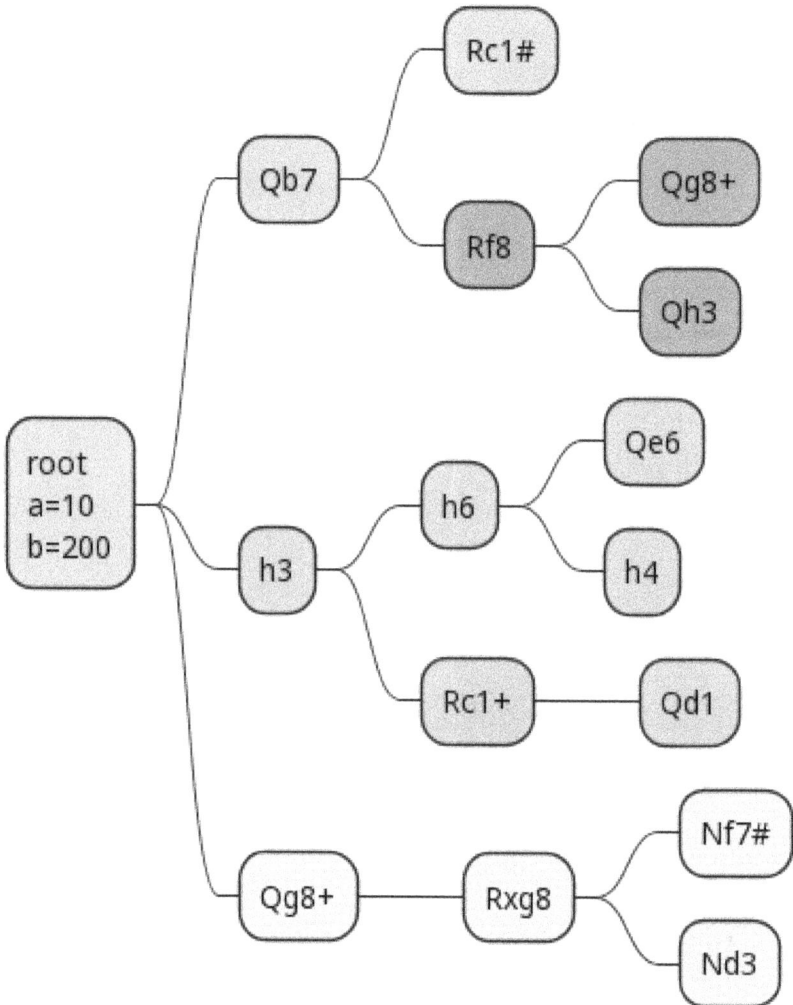

Now we make our move, and our opponent's reply is forced. If we happen to then also pick the right final move first, we prune out all the remaining

---

[8]https://www.chessprogramming.org/Quiescence_Search

ones.

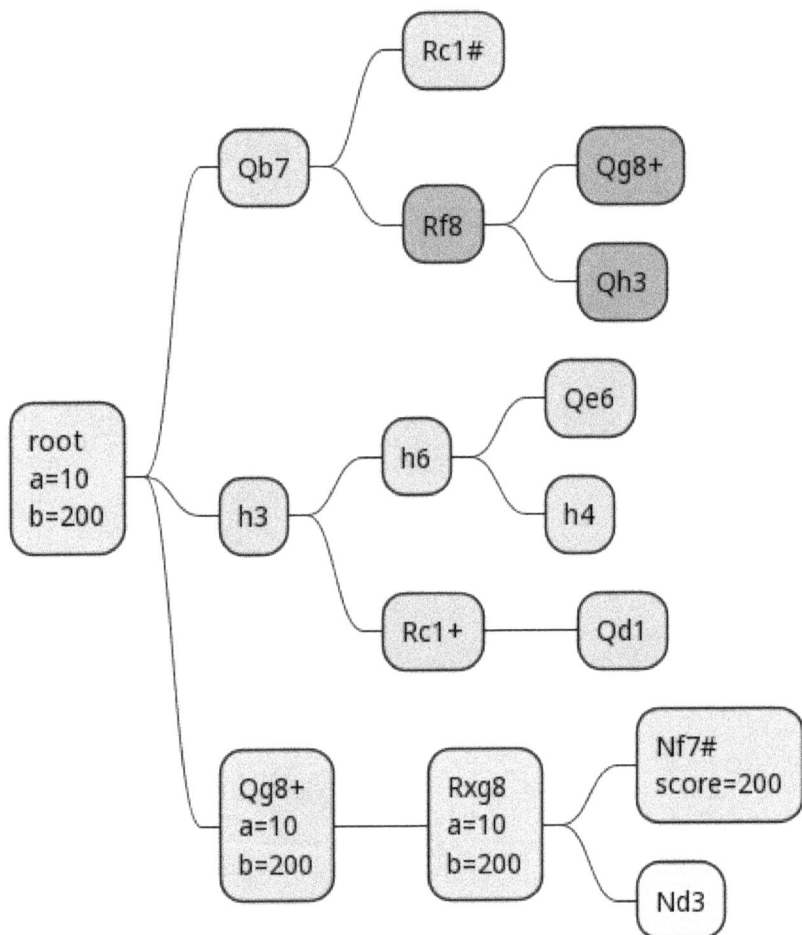

Working our way back up the search tree, we can see the effects on alpha and beta.

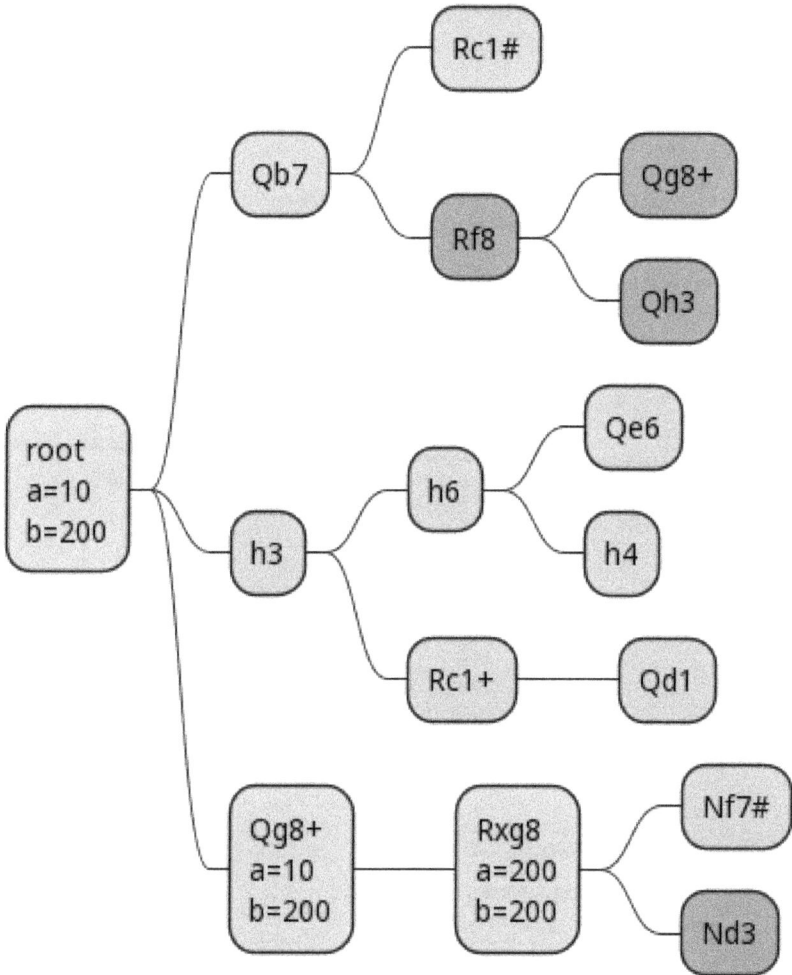

Once again we've found a situation where alpha = beta, so we can prune the rest of that tree. As we work our way back up, we eventually find that this is, indeed, the best move.

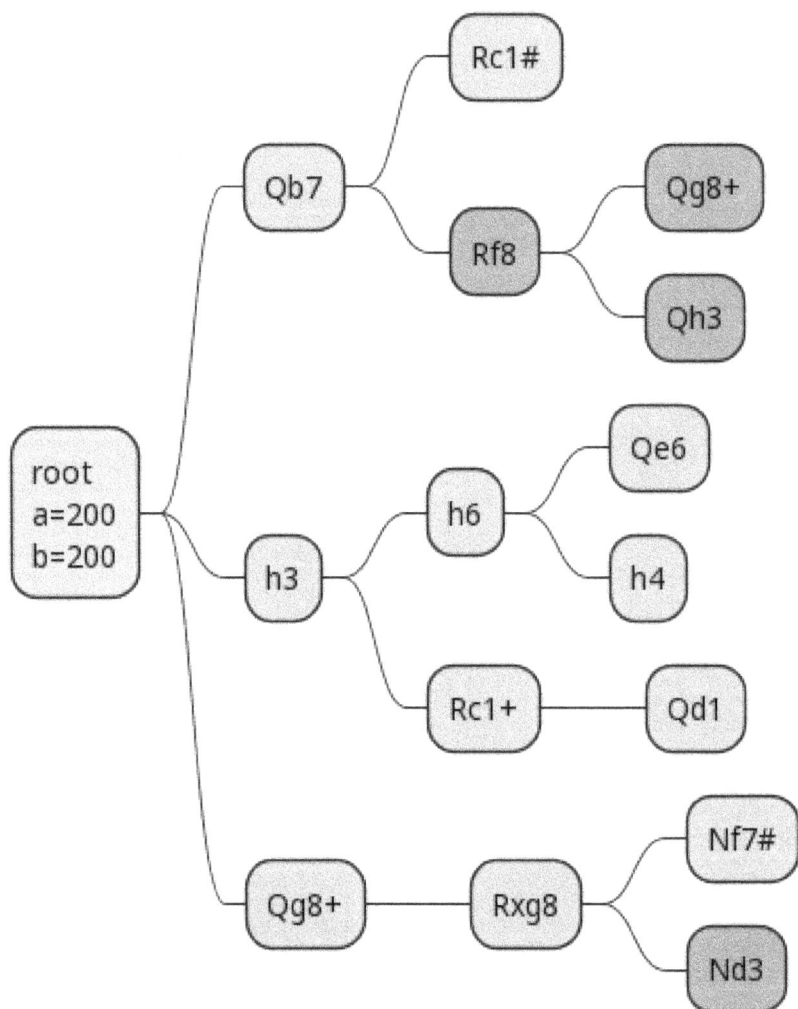

The evaluation for this position is, correctly, that white is going to win.

And the best part: with alpha-beta pruning, we only had to evaluate 4 leaf nodes, instead of 7!

12.4 I think I get it now!

This exercise was helpful for me in internalizing how alpha-beta pruning works. The fundamentals are pretty clear:

- At every round, you pick the move that maximizes your lower bound
- At every round, your opponent picks the move that minimizes your upper bound

Alpha is your lower bound, and beta is your upper bound.

Overall I get the algorithm better than I did before. It's still difficult for me to visualize and keep in my head, especially in the negamax form. Someday, I might make an interactive visualization for it, but not today!

One big takeaway from this is that alpha-beta pruning is making the same tradeoff we make in many systems: Increasing speed and efficiency at the cost of understandability. Most times when you optimize a program, unless you're swapping in a fundamentally more elegant solution, that optimization makes it harder to understand.

This isn't without risks! When it's harder to understand, it's easier to make mistakes and introduce bugs. We had a lot of bugs while implementing alpha-beta pruning the first time, and had to trace through it by hand. Having an easy visualizer for the search tree would have been helpful, but also has its limits: The whole search tree would be overwhelmingly large, and the parts that are helpful to trace are hard to pick without a human in the loop.

If anyone has good ideas on how to present the search tree to human users, I'd be all ears! I'd like Patzer to be at least somewhat comprehensible, and having nice visualizations on it would be a pretty cool angle on that.

13 RC Week 5: Wrapping up projects and starting a new one

Originally published on 2022.10.21.

Another week of my RC batch wraps up. I'm done with five weeks, and seven weeks are left! Time is still flying by, and I've hit an inflection point. I have gotten what I want out of the two projects I've worked on so far, so I'm going to wrap them up and move on to one new project for the rest of the batch.

13.1 A very social week

This week I did a lot of social things:

- 7 pairing sessions
- 8 coffee chats
- Went to an ML event
- Went to a theorem proving event (my brain is melted)

This was good, but I want to tone it down a *little* on the pairing and chats next week. I need to recharge, and I need to build a little more time for individual think time. Also for walks. Walks are nice.

13.2 Winding down my current projects

So far, I've worked on a key-value store[1] and a chess engine[2]. My goal with these was to learn about how systems programs are written in a way that's efficient and extensible. I think I've gotten there!

The key-value store supports a basic but useful subset of Redis commands, and its performance outpaces Redis (admittedly by using multiple threads, but single-thread performance nearly matches Redis even with write-ahead logging enabled). The next steps here would be to add more Redis commands or decide on some other more interesting features

---

[1] https://github.com/ntietz/anode-kv
[2] https://github.com/ntietz/patzer

to add. The performance constraints aren't particularly interesting; it's useful as-is if I made it complete and production ready, and I don't want to focus on making something production-ready.

And the chess engine is strong enough to beat me if I'm not *very* careful, although I can beat it if I pay attention and give it my best. The next steps here would be to add more things like quiescence search and iterative deepening. This is actually still very interesting to me! But I want to learn about data-intensive applications, and I can't focus on a new project while I keep working on the chess engine. I'm going to come back to the chess engine sometime, either at the end of my batch or after it. But I'm putting it on pause for now to give myself space to explore the new project.

13.3 The new project

I was toying with adding some features to the key-value store that would require some indexing, some disk read/writes, some user interactive queries. It all felt artificial, and it was a lot of design but in a space that had no real constraints, because there was no use case. Ultimately the goal isn't to add a feature to the key-value store. The goal is to learn about things like indexing, database query languages, data access patterns when it won't fit in memory.

What I need is a project that combines my learning goals with some practical, tangible problem that will impose constraints and give actual user requirements. I'm going to get that through creating a chess database. Sometimes "chess database" can mean "a bunch of PGN[3] files in a collection," but here I mean a database tailored to holding chess games and doing analysis on them.

The raw data that I have for master-level games is about 7 GB on disk. Lichess also has an open database of games[4], with each month of data being about 25 GB. If I want to do any sort of database analysis across play below the master level, this will quickly be larger than fits in RAM. (But I'm going to start with master-level play for now.)

I've created the project hub for IsabellaDB[5] (maybe I registered a domain as well, but let's not talk about that). If you're interested in pairing on it as it comes to live, reach out to me!

Some features I want to implement (some posed as questions I'd like to be able to answer):

- Standard chess database features:

---

[3]https://en.wikipedia.org/wiki/Portable_Game_Notation
[4]https://database.lichess.org/
[5]https://sr.ht/~ntietz/isabella-db/

- Explore openings and see what the win/lose/draw percentages are
- Detect when a given game has reached a novel position (eventually an integration into live broadcasts of games?)
- Filter/search by player name, rating, event, and other metadata

- For a given position, what other main-line positions can it transpose back into?
- If a player plays X opening as white, what do they typically play as black?
- For a given player, what is their repertoire? (potential integration with chess.com and lichess APIs to do some prep on real-life opponents!)
- Find the games in which there's a queen sacrifice
- Find positions where there is a battery, a pin, or another tactic available

    - Can this be used to construct tactics puzzles from real games?

I'm super excited to start working on this! I'm also a little overwhelmed, because it's a *lot*, and it's very much at the edge of my abilities right now. I went down a rabbit hole today on how to approach the UI, since a chess database needs something visual. Ultimately, on the advice of a couple of folks, I settled on just doing the most basic thing I can for the UI and then if I find more need for interactivity later, adding it on. Static results pages it is!

13.4 What's in store for next week?

Next week is all about shifting focus to my new project. I'm going to write up an initial plan, get some review on the plan, and start implementation!

- Keep pairing every day, keep coffee chats every day, but not too many days of doubling up!
- Write a blog post! I'm switching from GitHub to Sourcehut, and I want to talk about why
- Start on IsabellaDB[6]:

    - Write up a design for basic functionality: opening explorer, and searching games by metadata, and being able to scroll through the state of a game
    - Implement these features!
    - Pair with at least one person on implementing this stuff

[6]https://sr.ht/~ntietz/isabella-db/

- Read another Red Book[7] paper
- Read the first chapter of Software Foundations: Logical Foundations[8]. (We're switching to this instead of the Lean book. Hopefully brains are less melted.)
- Go to some of the other events, like creative coding and leetcode

It's going to be a full week, and I'm excited to get started. But first, it's time for a restful weekend. I'll be spending some time away from computers out in the workshop.

See you next week!

---

[7] http://redbook.io
[8] https://softwarefoundations.cis.upenn.edu/current/lf-current/index.html

14 Paper review: Concurrency Control Performance Modeling

Originally published on 2022.10.27.

Another week, another paper! This week for our Red Book[1] reading group, I read "Concurrency Control Performance Modeling"[2] by Rakesh Agrawal, Michael J. Carey, and Miron Livny. It was 46 pages, and I had a little trouble finding the whole paper—many of the Google Scholar links had missing pages in the middle, which was confusing the first time I encountered a weird gap.

My main takeaways from this paper were:

- The performance of a database is highly sensitive to its resources and workload in combination
- Simulation is a tremendously powerful technique for working on database performance and correctness (see also a fantastic talk about TigerBeetle[3] which also talks about simulations for DBs)
- When resources are limited, blocking transactions is probably better; when resources are unconstrained, restarts are probably better!
- There's some point of resource utilization below which, behavior resembles having infinite/unconstrained resources (this matches conventional wisdom where we see if disk/CPU goes above X% for our DB, time to make some changes)

Overall this paper was well written and easy to read. It used simulation and analysis to understand the behavior of databases under certain concurrency models. This worked better than simulation or analysis alone, and is definitely a technique to use in my own work.

Highly recommended reading if you're interested in understanding the performance of databases, as either a database engineer *or* as a user of databases. This paper will help you understand why, for example, having a ton of concurrent transactions in PostgreSQL can bring it to its knees, and you can get higher throughput by limiting the concurrency of connections.

---

[1] http://redbook.io

[2] https://scholar.google.com/scholar?hl=en&as_sdt=0%2C39&q=Concurrency+Control+Performance+Modeling%3A+Alternatives+and+Implications&btnG=

[3] https://www.youtube.com/watch?v=BH2jvJ74npM

14 *Paper review: Concurrency Control Performance Modeling*

This was a short review, because I don't think there's a whole lot else to say on the paper without directly repeating it. It uses a lot of charts, and the 46 pages are probably more like 20 pages of substance when you consider the space the charts take up, so it's a quick read.

15 My first impressions from a few weeks with Lean and Coq

Originally published on 2022.10.28.

For the last few weeks, some of us have been working through learning about interactive theorem proving together at Recurse Center. I've been curious about proof assistants since undergrad, and finally have the time, space, and peers to dive into it with. It's been an interesting experience getting started. Since we're just getting started, I can't tell you much about the long-term experience, but I can give some basic guidance on what it's like to get started on each and who I imagine the audience for each is.

First off, **what's a proof assistant?** Simply, it's a piece of software that helps develop formal proofs via human-machine collaboration. You want formal proofs in a lot of cases; they're used in math, but it would also be great to know that an algorithm you want to implement does what you say it does, or that a bigger piece of software is proven correct. Proving software correct does, of course, lead to the question of how you check that the spec is correct, and that you're specifying the properties you care about. That's a whole other conversation.

There are a bunch of different proof assistants available. The big names are: - Coq[1] - Lean[2] - Isabelle[3]

From the outside it's hard to know which one to pick based on features of the proof assistant itself, so we chose based on the material available to learn from. If you learn one, it'll make learning the others easier, so going off what's most accessible to learn is a great approach.

Where we went wrong is we *thought* we picked the one that was most accessible to learn. We started out with Lean, using Theorem Proving in Lean 4[4]. When we started, I wasn't aware that there's a split in the community. The maintainers of Lean have moved on from Lean 3 and started Lean 4, which aims to (among other things) also be a fully-featured general purpose programming language. Unfortunately, much of the material out there is for Lean 3 (such as the impressive library of math that

---

[1] https://en.wikipedia.org/wiki/Coq
[2] https://en.wikipedia.org/wiki/Lean_(proof_assistant)
[3] https://en.wikipedia.org/wiki/Lean_(proof_assistant)
[4] https://leanprover.github.io/theorem_proving_in_lean4/

they're proving as a community!) which is now also maintained by the community in a fork[5]. I'm not sure which would be better to learn. If I tried it again, I'd probably try Lean 3 since the community is there, but I'd also probably not try again.

I did take away some important concepts from our brief misadventures with lean, and there were some positives. The tooling was very nice and easy to install. Lean has a nice LSP[6] implementation, making it easy to integrate with your text editor of choice, and there are robust plugins available. The whole thing was nice and easy to install. But that's sort of where the fun ended.

The learning material was very choppy and difficult for us to work through. It had sparse exercises, and there was a sudden cliff of complexity in the first chapter. Overall it felt like the target audience (appropriately) was mathematicians, and we're decidedly *not* mathematicians.

We moved on and started reading Logical Foundations[7], the first book in the Software Foundations series. This series uses Coq, and it is targeting folks who are specifically interested in software, not in math. For people who want to learn proof assistants with a software background, this feels like **a much better choice**. It also helps that this is written by a group of professors who have a wealth of teaching experience, and it comes bundled with both exercises and an autograder for some of those exercises, so it's feasible to work through without an instructor.

The first chapter of Logical Foundations went much better for us than the first chapter of the Lean book. There were things we didn't understand right away, and things we had to work through as a group. In my opinion, this means that we found something that's the right level of difficulty: It wasn't so hard that we can't get through it (hi, Lean), nor was it so easy that we're not learning. It's a difficult that feels achievable but definitely stretches us.

And that's right on brand for working through something at Recurse Center, where one of the main principles is to work at the edge of your abilities.

Coq itself was not without its difficulties, however. In particular, one of my fellow Recursers had a non-trivial time getting it installed. This might be a particular issue with M1 Mac support, because I had a package available for nice and easy installation on Fedora. It wasn't as easy setup as Lean, but then it eventually got out of our way.

In retrospect, Coq is also a much more solid choice for us to learn, curriculum and tooling aside. It sees more use in the software industry than

---

[5]https://github.com/leanprover-community/lean
[6]https://en.wikipedia.org/wiki/Language_Server_Protocol
[7]https://softwarefoundations.cis.upenn.edu/current/lf-current/index.html

Lean, and has been used to produce CompCert[8], a C compiler which has been formally proven. (Not that I'm jumping to use CompCert: I'd still be writing C, and my *own* programs would be riddled with memory errors.) Isabelle is also a solid choice. It's used to verify the seL4 Microkernel[9], and Martin Kleppmann has used it to verify distributed algorithms[10]. We didn't choose simply because we found good resources for Coq but not Isabelle. I'd like to explore Isabelle someday, because it looks a little more explicit than Coq, which I think would be more to my taste. If you know any Isabelle resources, please send them my way!

Overall it's been a pretty great experience learning a proof assistant, in large part due to having peers learning it with me. (Shoutout to Mary, Paul, and Ed!) I'd highly recommend trying it out if you're interested. It's less scary than it seems—if you have the right material to learn from.

---

[8] https://compcert.org/doc/index.html
[9] https://sel4.systems/
[10] https://martin.kleppmann.com/2022/10/12/verifying-distributed-systems-isabelle.html

16 RC Week 6: Halfway done, wrote a parser!

Originally published on 2022.10.29.

I'm halfway done with my RC batch now. Time feels like it has sped up. The feeling that my time at RC is infinite is gone. This was compounded by seeing folks from the Fall 1 batch conclude their batches yesterday. We'll get a new boost from the Winter 1 batch joining on Monday, which I'm really pumped for! New people, new excitement, new energy!

I'm happy with how things have gone so far in the batch. I don't think I want to do anything dramatically different in the second half of the batch, except be a little bit more focused on one project instead of splitting between two.

I did have a less social week this week than most weeks, because I have some personal life stress right now (should wrap up next week) and it made it hard to focus, and I withdrew a little bit. Despite that, I still had a pretty social week! Something for me to take away here is that RC has shifted my understanding of where I get energy from and how much I do benefit from social things.

This week, I... - had 5 pairing sessions - had 5 coffee chats - went to the Rust and theorem prover groups

Next week I'm going to have more coffee chats probably, since the new folks are joining. But I'm going to remember to be kind to myself, and to be realistic. There are some factors outside of my control (taking a family member to a couple of appointments, plus two 8-hour drives to/from there) which will limit how much I can do next week.

But! I did get through some code this week, and I really want to share what I did.

16.1 Wrote my first parser!

I learned to use the nom[1], a parser combinator library. I wrote a PGN[2] parser (and improved it and sped it up, with the help of two other Recursers, yay pairing and community!) which worked pretty well. It can parse

---

[1] https://github.com/Geal/nom/
[2] https://en.wikipedia.org/wiki/Portable_Game_Notation

16.3 What's planned next for IsabellaDB?

The data storage and retrieval side of things is a little fuzzy for me until I get in the weeds, just some details are out of focus. But I think the thing that's the biggest unknown really comes down to product type things:

- What queries will users want to do? (This is needed to choose what things to index on!)
- How should users interact with it? What should the query language be like?

Because these unknowns are product-y, I'm focusing right now on getting something usable that I can start playing with.

Next week, I want to get the position index up (so I can, given a position, find all other games that contained that position) and build a UI that exposes searching positions by clicking through an opening tree. That'll be enough for me to start thinking about how I want to use it. I'm also going to continue pondering design: I have some ideas on how to pull out fields to a columnar store, but I'm less clear on how query planning and optimization will work in *any* format, so that's on my docket to learn about!

Next week I'll have a pretty full schedule:

- Lighter pairing load, because of personal life obligations
- Coffee chats at least once a day! It'll be exciting with the new batch coming in.
- Write a couple of blog posts. I have two that are in the queue that I have some research done for, so I need to buckle down and write them. I'll stagger their releases.
- Work on IsabellaDB (position index and frontend)
- Finish reading and summarizing my current Red Book paper

It's going to be a full week! I'm excited to welcome in the new batch, and keep in contact with all the folks from Fall 1.

See you next week!

17 RC Week 7: Four habits to improve as a programmer

Originally published on 2022.11.04.

Seven weeks down, five weeks to go! It's flying by quickly. On the one hand, I want it to last forever. On the other hand, I know it can't, and I'm looking forward to talking to coworkers again at my day job when I go back. RC has given me a renewed appreciation for what I get at my job. More on that in December, though.

For now: RC, and the goals I have while I'm here. Instead of focusing on what I did this week, I'm going to use this post to talk about how I want to improve as a programmer.

I came into RC in with certain hard skills I wanted to improve (debugging! profiling! optimization!) and those are all great, and I *have* worked on them. But I am starting to focus more on the meta level of how do I actually improve broadly as a programmer? What habits do I have that make me less effective? What habits do I have that make me *more* effective?

One of my bad habits is jumping into writing code without doing sufficient analysis. This might surprise some of my coworkers, since I'm the design doc person at work. I **love** writing design docs. But what I don't do as great a job of is doing analysis when I'm midstream on something.

Last week, I wrote a parser. When I was working on it, I ran into some significant memory consumption issues. I was pairing with another recurser, Manuel, and I started to just kind poke around at different ideas to bring down memory consumption. Instead, he pushed us to take a few minutes to actually calculate what a lower bound is on the memory which would be used. This informed our approach for reducing its usage and figuring out where high usage was coming from, and ultimately was a lot more effective a lot more quickly.

So that's the first habit I want to build: **Using estimation to inform how I work on things.** It sounds simple, and it's hard for me to break out of the flow of coding to do it. But I've already started to put it in practice this week, which is promising!

The second bad habit I have is not reading the docs. This one is pretty self-explanatory. I want to read the docs more when I start using a new library. But I also want to read the docs for standard library things that I *don't* use often so that I know they're there when I need them. I might need to make some Anki cards to study the standard library or something!

And that's the second habit I want to build: **Proactively read/study documentation.**

But what habits do I have that make me a better programmer, that I'm already doing?

The main one is *writing*. I believe I'm at least a halfway-decent writer[1]. The thing that I do that makes me more effective is **writing design docs**.

For the first half of RC, I didn't do this, and I'm not sure why. I've started writing design docs[2] for IsabellaDB and it's been so helpful in clarifying my thoughts and giving me discrete steps that are more approachable. My designs are not correct on the first pass, but that's also kind of the point! After this database is working, I'll have a record of the design iterations, which I think will be a kind of neat example of how a skilled software engineer goes through design iterations, especially in an area they're not highly familiar with.

So that's the third habit, which I want to reinforce: **Write design docs proactively and often.**

I've also fallen out of the habit of using my favorite and most effective programming tool: Walking or running. Something about walking or running helps me work through problems that have been nagging me. The work I'm most proud of in my life has all been catalyzed by walks or runs. I don't solve the entire thing, but I typically will get the kernel of a solution that I need in order to chip away at it. I used to go for walks proactively during the day to think, and I've fallen out of that habit.

So that's the fourth habit, which I'm going to rebuild: **Go for walks during the day.**

These habits are what are effective *for me*. I'm going to keep at it. I'd be eager to hear from any of you: What habits have you found make **you** a more effective programmer? And are there any you think would help me that I'm missing?

---

[1]Software engineers are often also *not* good writers, so this may play to my advantage. I don't know how we as a profession compare to the broader population, though. I'd be curious to learn if there's any work done comparing the writing abilities of different professions!

[2]https://git.sr.ht/~ntietz/isabella-db/tree/main/item/docs

18 Paper review: C-store

Originally published on 2022.11.04.

It's that time again: I read another paper, and here's what I took away from it! This week I read "C-store: a column-oriented DBMS" from chapter 4 of the Red Book[1]. This one I picked since I thought it would be helpful for the chess database I'm working on, and it does seem applicable!

This paper was pretty significant for making a strong case for the utility of columnar databases in read-heavy situations. It demonstrated an architecture for a column database which not only beats row-based databases of the time (in their workload) but also *beat the proprietary column databases* of the time as well. One of their key takeaways is that being columnar gives you: - Very good compression that's not feasible with row stores - A reduction in overhead of storing records - The ability to have multiple sorted orders for a column for efficient querying

The overall architecture they presented seems straightforward and perhaps deceptively simple. They have three major components: - WS, the writeable store - RS, the readable store - Tuple mover, to move written data from WS into RS in bulk periodically

Each of these components was described in brief detail. There was enough detail to get the gist, but not enough to go write an implementation myself. I think this is the nature of publishing: You have limited space to publish in, and also it would be nice to save some details to publish later. They also have a number of things which were planned but not implemented, so sparse detail may also be from simply not having answers.

Some of the things I was left wondering were: - When does the tuple mover decide that data is suitable to merge? - How does the tuple mover merge process work? - What ratio of reads:writes does this architecture support? Beyond what point does the WS become a bottleneck? - What workloads are favorable to the row stores over the column store? - How are the projections chosen? (This last one is probably my biggest open question.)

This paper really gave me some inspiration for how to structure the database I'm working on. Hopefully I'll have a post up about that database's structure once it's working (always more to do, always

---

[1] http://www.redbook.io/ch4-newdbms.html

performance traps to fall in before it's done!) and I'll be able to talk more about how its design was informed by C-store.

Next week's paper will be the DynamoDB paper, which I'm excited to read! Later!

19 Open source licenses as a reflection of values

Originally published on 2022.11.08.

I'm the kind of nerd that has a favorite software license. For a while that favorite license was the Mozilla Public License[1] (MPL). Right now, it's the GNU Affero General Public License[2] (AGPL). Licenses are really important on all code, and they're critical to the open source movement. They reflect the values of the people writing the software. Let's talk about software licenses and why they matter, then how they reflect our values. (And of course: I'm not a lawyer, and there may be errors in this!)

19.1 The importance of licenses

The first question is: Why do licenses exist and why do they matter? I'll take a US-centric view here, because that's what I'm most familiar with.

In the US, all code is by default protected by copyright, both as the source code and in compiled form[3]. This means that other people don't have the right to use your code (with some possible exceptions) without permission. Software is less useful without users (as are books without readers, etc.) so we want some way to let people use our software. That's where copyright assignment and licenses come in.

If you develop code for your employer, they probably require you to assign copyright to them for the code you write. There are some notable exceptions[4], but this is common in US employment contracts so that your employer owns the rights to the code. This lets them choose the license for the code, sublicense it, sell it—all the things a business typically wants to be able to do.

If you're developing code outside of employment, and you retain the copyright on that code, then you get into license territory. When you put up code for other people to see, if you don't include a license, they have **no rights to use that code**. You have to include a license for people to be able to lawfully use, copy, modify, or distribute the code.

---

[1] https://www.mozilla.org/en-US/MPL/
[2] https://www.gnu.org/licenses/agpl-3.0.en.html
[3] Code developed by the US Government *cannot* be copyrighted and enters the public domain.
[4] https://sourcehut.org/blog/2022-10-09-ip-assignment-or-lack-thereof/

So if you are publishing code on the Internet, it's best to put a license on it! Otherwise, it might solve someone's problem now or in a decade, but they could be out of luck, unable to actually use it.

19.2 Open source and licenses

Licenses are especially important if you're writing open source code. Open source requires more than making the source code available to view. The full definition[5] by the OSI specifies ten requirements for being considered open source. There are many licenses which satisfy these requirements, and they include the ability to freely redistribute and modify the code and derived works.

Some of the common licenses are: MIT, BSD, GPL (LGPL, AGPL), Apache, and MPL. You'll also see some licenses like the Server Side Public License[6] floating around which capture some but not all of the definition of open source; notably, they restrict commercial activity to try to prevent companies like Amazon from squashing other companies like Elastic and MongoDB (which are themselves quite large now: Elastic and MongoDB have market caps of $6 billion and $12 billion respectively).

With the myriad licenses out there, it's helpful to classify them. There are a lot of little subtle differences (trademark use, patent rights, etc.) but the biggest differentiation is **permissive** vs. **copyleft** licenses.

A **permissive** open source license is one where almost all rights are granted on the software, including the right to relicense it and even make a closed-source fork of the software. This category includes the Apache, MIT, and BSD licenses. It's common to see libraries licensed under this, because then almost any developer can use your library: There won't be license concerns (maybe patent or trademark concerns). Among these, the MIT license is the most common. Many Rust crates are dual-licensed under MIT and Apache.

In contrast, a **copyleft** open source license restricts more rights on the downstream recipient of the software because it requires that any modifications or redistribution of the software are released under *the same terms* as the original software. That is, if you distribute a modified copy of a copyleft program, you must make that modified copy available under the same license as the original program. This category includes the GPL and its derivatives and the MPL. The most widely used piece of software in this category is probably the Linux kernel, a little project a few people have used. Many open-source applications and libraries use this

---

[5]https://opensource.org/osd
[6]https://en.wikipedia.org/wiki/Server_Side_Public_License

license to ensure that contributors contribute their patches back to the community.

The main differences between the copyleft licenses boil down to a few points. GPL[7] is the base case copyleft license to start from. If you use GPL-licensed in your program, your program is now a derivative work and must also be released under the GPL. This was written before web applications were the main form of distribution, so notably if you run GPL software on a web server it's not clear that that does constitute distribution. Comparing to GPL, the other common copyleft licenses have these differences: - AGPL[8] explicitly states that use over a network is distribution, so if a web application uses AGPL code, then it must also be released under AGPL - MPL[9] is licensed on a per-file basis, so it **doesn't spread** to the rest of your program; only changes to MPL-licensed files have to be released - LGPL[10] allows dynamic linking without the rest of the program being LGPL licensed, but there's nuance here

19.3 Reflections of values

Why are there two different major camps of open-source licenses? If we all have this shared belief that open-sourcing code is a positive thing, why isn't there a standard single license?

Because each of these licenses reflects subtly different values. Yes, we value other people being able to use and modify our source code. That's the base common value for the open source movement.

But beyond that, there are disagreements. Permissive licenses put high value on **full freedom** and there's higher value placed on people being able to create *anything they want* from the software, including proprietary forks, than on the open-source aspect itself. In contrast, copyleft licenses put high value on **preserving open-source**, and place higher value on keeping derivative works open than on fully free derivative use.

Other licenses also reflect values. "Source-available" licenses, like the SSPL, are increasingly prevalent. One pattern that we're seeing (at MongoDB, Elastic, and LightBend / Akka, among others) is the relicensing of open-source code from a copyleft license to a source-available license. In my opinion, this is *theft from the community*. (This is only possible if contributors assign copyright to the entity behind the project.) This relicensing explicitly says that **building a massive business** is more

---

[7]https://choosealicense.com/licenses/gpl-3.0/
[8]https://choosealicense.com/licenses/agpl-3.0/
[9]https://choosealicense.com/licenses/mpl-2.0/
[10]https://choosealicense.com/licenses/lgpl-3.0/

valuable than the community benefiting from their (volunteer!) contributions.

19.4 How I learned to love the AGPL

When I started in the software industry, I did not develop *any* open source code. I had some config files up on GitHub which were largely not copyrightable, and people could draw inspiration from them. Everything else I did, I kept private and did not publish the source code, let alone license it for others to use. This was in large part driven by two things: - a **fear of publishing "bad" code**, and - a **delusion that I could become rich** by keeping my software closed.

The first point is somewhat self-explanatory. I was afraid, and it felt like all the eyeballs of the world could be on me if I put my code out there. Would someone think I'm dumb if I publish my code??? (Probably not, and if they do: their problem, not mine.)

The second one is a *little* more subtle. I entered college right after the 2009 financial crisis[11], so as I entered college everything had hit rock bottom. Some of my friends had to change where they were going to school when their first choice colleges had to rescind scholarships due to collapsing values of endowments. It was a wild time, but hitting bottom meant that the market was only going up from there. So during college, I saw startups being formed, hitting huge valuations, growing massively. The companies that were founded before 2009 and survived became incredibly valuable. Why not go for that? Money seems nice.

So with that approach, I published almost no open source code. A few things were open, like my config files and some code from my college classes. The rest I kept private. I felt that the main value I was creating with code was **potential economic value**.

Eventually, I started to open-source more things. It took a long, long time to get comfortable putting anything out there. Once I did start putting things out there, I had two licenses I tended to use: - GPL for things that were just for sharing, like my Advent of Code solutions - MPL for things that I thought could someday be a Big Deal and wanted to have rights over

I chose GPL for the things that were just for sharing because I felt vaguely uncomfortable with **other** people having the potential of using my not-quite-throwaway but not-quite-good code for something profitable. On the other hand, I used MPL for things where I thought they could become big. I picked it because it lets you use an open-core model: Since the

---

[11] https://en.wikipedia.org/wiki/Global_financial_crisis_in_2009

license applies on a per-file basis, it's feasible to combine an MPL core with proprietary extensions. This was driven by a desire for *me* to be able to profit from the work I was doing if it became a big deal.

That's delusional thinking, though. Unless you put in a lot of effort to getting something to be big, it's exceedingly unlikely that a community will just **poof** into existence around it. And it's very unlikely that it'll become something valuable to commercialize without that effort, as well. Commercializing isn't something that **happens by chance**. It takes deliberate effort and planning, and that can distract from the goal.

I didn't start any of those projects in order to get rich or to make a commercial project, but that thought in the back of my mind limited my license choice. Joining Recurse Center[12] gave me the space to remember again why I start projects and why I write code in the first place. Yes, I enjoy making a great living and having a great paycheck. No question. But I got into this field because while I love the creative aspect of writing programs.

Something clicked over the past 7 weeks at RC, and I remembered why I'm doing all this. Or rather, I remembered the reasons that don't motivate me. Building a billion dollar business is definitively **not the motivation**. And making my code available so that other people can make billion dollar businesses on top of it? Nah. Fuck that.

From here on out, if I release open source software written on my own time, it's very likely going to be under the AGPL. I don't want people to be able to use my code in proprietary software unless they paid me to write that code. I don't want to make a proprietary version of my own software, either! If I write a useful library, I would love if other copyleft projects can use it. And I rather do not want it to contribute to the value of a big business without them giving back to the community. I just want it to be open and free.

"But what if it's something really valuable, and now a company can't use my library?" That's great! That's the idea!

For me, this clarifies why I'm writing code: The value is *the software itself*. The community, if one arises, owns the code.

Copyleft all the way!

19.5 What's the right license for *your* project?

To pick a license you have to first figure out what you value and what the goals for the project are. Only you can answer what your primary values

---

[12] https://www.recurse.com/

are and why you're writing the software. Ultimately, the **license is a reflection of your values**, so think about what you value before you pick what license to use.

If your primary value is **open-source itself** and you want **community ownership**, copyleft is a clear choice. There are very few downsides to using a copyleft license like the AGPL for your code. The downside is simply that you cannot make a proprietary work from your code. The upside is that no one cannot make a proprietary work from your code[13] (but businesses can be made[14]).

By using AGPL for your libraries, you can use almost any other library in your own. You have few limitations on what you can pull in, so you can make full use of the open source ecosystem.

That said, there are plenty of situations where you would want to use a non-copyleft license.

Here are a few examples: - You simply want anyone to be able to use it however they want, and you don't care if derivatives are open-source. This is saying "I have this cool thing, use it however you like!" People may learn from it, they may make a product from it, they may contribute back to it! It's all cool! - You're working on something that is to be consumed by other businesses. If you're working on a developer tool, for example, and you want widespread adoption of it in businesses, copyleft licenses can pose serious problems. Even if it would pass legal muster, businesses are cautious with licenses. Using a permissive license can make it easier to get widespread adoption since people can use it to achieve their day jobs and to write other non-copyleft software. - You're working on something where you explicitly *do* want to empower proprietary software creation. We live in a capitalist society, and you may want to empower business creation. Using a permissive license can remove a lot of friction from this, if your goal is creating economic value.

When I chose licenses that permitted proprietary software creation, that was a reflection of my values. My shift to using AGPL for all the things is a reflection in a realignment of my values.

So what do you value? Pick a license that displays that. And please consider valuing open source and community by choosing copyleft licenses.

---

[13]This does require license enforcement if there are violations. I have no familiarity with this, and it intimidates me.

[14]Note that there is a distinction here between making a proprietary work (where they own all the rights) and building a business on top of it. You can absolutely build businesses on open-source projects through things like support contracts and hosting services. I do think this is a harder path than building a business around proprietary software, but copyleft licenses do not *preclude* commercial activity.

Huge thanks to Julia, Joe, and Ed for giving me feedback on this post before publication!

---

20 RC Week 8: Life happens, and databases are hard

Originally published on 2022.11.12.

I'm two-thirds of the way done with my RC batch now. Eight weeks down, four weeks to go.

The last two weeks have been difficult for me because of life happening. Week 7 was hard because I had some travel to help my parents, and that just takes me out of my routine and is generally stressful. It was good, and I am very glad that I had the opportunity to help. But it was still a lot, and it made it hard to make significant progress on my project. And week 8 was hard because I finally caught the cold that's been going through my household. My brain was just... fuzzy this week.

In that vein, I spent the last two weeks mostly focused on figuring out how to build an inverted index over all the unique positions in a collection of chess games I have. To make it concrete, this is 3.8 million games and about 240 million unique positions (north of 300 million before dedup). I ran in circles, and I think it was a combination of: - Life happened so I wasn't at my best - Databases are hard

I'm not upset, though! By trying a lot of dead-ends, I got to understand the problem space more deeply. For example, I learned that:

- Chess positions cannot be represented in a fixed-size struct of less than ~29 bytes[1] (and that is a *maybe*)
- Rust's HashMap implementation is based on a hashmap created at Google (video[2] explaining how it works)
- Data-oriented design[3] gives many practical benefits in structuring data to reduce overhead, such as storing a struct of lists instead of a list of structs
- 64-bit hashes can handle a *lot* of elements (4 billion-ish?) before you expect to see your first collision

---

[1]This blog post used to say that it couldn't be less than 36 bytes. I think that *might* be true if you use 1 byte per piece and then bit-pack the rest of the information, but a fellow Recurser and I worked out that it can indeed be smaller. Right now speculatively we think it can get down to 29 bytes, but I'm not about to write an implementation to prove it.

[2]https://www.youtube.com/watch?app=desktop&v=ncHmEUmJZf4

[3]https://vimeo.com/649009599

So after these last two weeks, I've finally gotten the index built! And it saves it out to the disk, which I can load back in to quickly find the games which contain a given position. (How quickly is yet to be seen— I'll benchmark it, and have a few ideas for improvements if it's slower than necessary.)

Now I'm moving forward on building an application on top of this index. I'm going to first make an opening tree explorer, where you can click through from the beginning of a game and see how many games occurred with that position, the results from there, and drill into a (partial) list of the actual games containing that position. This will require building out a basic frontend (entirely HTML/CSS for now! I don't think this needs much, if any, JS), and it will also require adding some additional basic indexes, like bitmaps of game results.

Next week, I'm hoping to have something that I can demo! It will be rough-and-ready, but it'll be the start, and then I can spend a few more weeks adding in more interesting query support and more filters on the games. Long-term, I think that isabella-db[4] can fill a gap in chess tooling today by making it possible to query for really interesting sequences of positions in games, like where sacrifices occur or where tactics are available. (This will likely also require integrating with an engine!)

I want to get more folks involved in this project, and the sooner the better. If you're interested in **being an alpha user** or **helping with the queries and indexing**, please reach out to me by email (or on Zulip, if you're a Recurser). I'm excited to see what I learn via this project for the rest of my batch, and where it goes after that!

See you all next week!

---

[4]https://sr.ht/~ntietz/isabella-db/

21 I'm moving my projects off GitHub

Originally published on 2022.11.16.

It's time for me to leave GitHub behind and move to another forge. I'm not necessarily advocating for anyone else to do the same, but if my reasons resonate with you then you may want to consider it. I also don't expect this post to... matter, if that makes sense[1]. I'm not a major open-source maintainer or contributor. I'm just somebody who likes to write code and likes to put it out there.

So, why am I moving my projects off of GitHub?

21.1 My Issues with GitHub

It ultimately comes down to some issues I have with GitHub, both as a product and philosophically.

The tangible one that tipped me to finally move: **I'm upset about GitHub Copilot.** It's fairly well known that Copilot can reproduce significant pieces of open-source code, stripped of their license[2]. I'm moving to make it a **little bit harder** to have Copilot train on my code. This is perhaps a futile protest, but it's what I can do an individual. Writing about this is another aspect of what I can do.

I hope that the ongoing litigation[3] gets us some clarity in what is legal here. In the US, a lot of "is this legal" is deferred to the run-time execution of contracts by courts, so this is our chance to find out what is legal or not. Hopefully this goes in the direction of defending open source and requiring attribution and copyright. As it stands, Microsoft/GitHub have basically washed their hands of it for users, saying reproduction of code is rare and the users must make sure it doesn't happen. Which... they're supposed to do how, exactly?

---

[1]So why write a blog post if I don't think it'll "matter"? Because this is **my** blog, and I write what I want! I'm not writing to achieve any goal or effect change. I do think it will be interesting to someone else, and writing publicly is a hobby I enjoy.

[2]See Drew Devault's post "GitHub Copilot and open source laundering" (https://drewdevault.com/2022/06/23/Copilot-GPL-washing.html) for a good post on this topic.

[3]https://githubcopilotlitigation.com/

Anyway, I don't want this post to be a full-on rant about just Copilot (I've got plenty of *other* things to rant about ☺) Copilot was just the tipping point. There are plenty of other issues I have with GitHub which are more significant for the decision to leave.

First, I think that **open-source code should use an open-source forge**. It's deeply ironic that the biggest forge for open-source code is itself proprietary. (And ironic that one of its biggest competitors, GitLab, is open-source and hosts tons of proprietary code.) I think this is not healthy for open-source in the long term. It gives a *lot* of control over open-source to Microsoft, and concentrating that power in one entity is not good, regardless of who that one entity is. I think they've mostly acted as good stewards so far, and this is about mitigating a risk and addressing a philosophical issue.

I also don't like how **GitHub changes my behavior**. This one is somewhat on me (personal responsibility) but this also comes down to how many modern tools are designed. Modern web design leverages and exploits human psychology to achieve the outcomes it wants (ultimately, increasing revenue, usually by driving usage). GitHub in particular is pretty effective at doling out dopamine hits to me. As a GitHub user, I was always seeking green squares to try to make sure I had activity every day. This led me to change my workflow to generate more green squares, not for whatever is maximally effective. Having visible stars and followers also turns it into a sort of popularity contest. There's a dopamine hit when you get one of those, so it creates a strong reward function for attention-seeking behavior.

Tools should be designed to help the user, not to help the company[4]. Many of GitHub's features can be defended (and I'm sure readers of this post will do so!). They certainly don't work for me, though. Using GitHub changes my behavior in a way that, ultimately, I find to be negative. So: bye, GitHub!

Another small reason is that I believe in **paying for my tools**. This is why I pay for email with Fastmail instead of using Gmail. The incentives are clearer when you're a paying customer than when you're a free user. I mean, they have to make money off of you somehow. The innocent explanation for free usage is as a loss leader to funnel people into an eventual enterprise sales cycle. The cynical explanation is to take more control over open-source and also to make a massive dataset to power Copilot and other products. If you're not paying, you're the product, after all.

---

[4]This is another reason I want to be on a FOSS forge. Incentives are better aligned when there is less profit motive and better philosophical alignment. Everything boils down to behavior and incentives.

21.2 Why I Chose Sourcehut

After deciding to leave GitHub, I had to pick a new forge. I settled on Sourcehut[5] for a variety of reasons.

The non-negotiable criterion was that **it's open-source**. The platform itself is licensed under the AGPL (mostly) and you can self-host it. They also provide consulting on open-source projects to get some revenue, and they don't require copyright assignment for contributions to Sourcehut from volunteers or employees. All of this is pretty strongly aligned with my philosophy and I appreciate it.

Another big reason is that Sourcehut is **explicitly designed to *not* dispense dopamine**. My brain's response to dopamine is widely exploited by our industry, which is why I'm not on any social media anymore. GitHub dispenses a *lot* of dopamine and it makes me change how I work to get those nice little green squares. Sourcehut rejects features that are only for dopamine hits without utility on their own, and is generally designed in a humane way that doesn't exploit human psychology. This makes my life tangibly better.

I also feel like **the platform's direction is understandable**. Some people criticize the maintainer (Drew DeVault) for having strong opinions. He does have those, and Sourcehut reflects it. This lets you have a pretty good understanding of where things are going and what he won't compromise on. In contrast, with a proprietary platform like GitHub, you can't quite be sure of the long-term direction. It depends on the company strategy and what metrics they're trying to optimize and who's making design decisions. Drew is transparent with things and even though he has strong opinions, it's a big tent. Unless you're working on blockchain projects that waste energy for imaginary magic beans. Then you can get out of this tent—and indeed, your data is portable, and you can easily migrate off to somewhere else!

Sourcehut is also **very light** and aesthetically pleasing. (I know aesthetics are relative.) It uses no JavaScript and page loads are just *wicked* fast. I've long bemoaned how everything is a SPA these days, and Sourcehut is as nice reversal of that.

As an example of how much lighter it is, let's look at one of my repos on both platforms. - On GitHub, my config repo[6] takes 933 ms to load and downloads 2.5 MB. - On Sourcehut, my config repo[7] takes 158 ms to load and downloads 138 kB.

---

[5] https://sourcehut.org/
[6] https://github.com/ntietz/config
[7] https://git.sr.ht/~ntietz/config

One aspect of forges today is CI or build tool. Having CI on the platform was an important part of me moving to Sourcehut (where Gitea, for instance, doesn't have CI built in). The delightful surprise was **how *good* Sourcehut's builds are**. I find GitHub Actions fairly confusing and difficult to get working. It feels like whenever I'm updating a GitHub Action, I have a string of ten or more commits that are all of the form "whoops, now does it work?". The only way to update it is to keep pushing to the repo, so you're left with a string of ugly commits while you iterate on your CI. Maybe I'm just dense, but this is an experience I've heard from others as well.

In contrast, Sourcehut's builds are quite easy for me to understand and use. You write a *minimal* YAML file describing the build (which OS, what packages you want installed, and a list of shell scripts to run) and then you get a VM that is running that for you! The model itself is pretty easy to understand, and the absolute delight is **how easy it is to debug builds**. With GitHub Actions, you keep pushing commits to debug. With Sourcehut Builds, you can do that, but you have two more powerful options: - **You can run ad hoc builds.** This lets you keep iterating on a build and test it before you commit at *all*, which keeps your commit history clean and is a much nicer workflow. - **You can SSH into your build VM.** If you have a failed build, the VM sticks around for a bit and you can connect in to try things and figure out what you should change to make the build work next time. This is such a helpful tool.

The elephant in the room with Sourcehut is, of course, its **contribution workflow**. Projects on Sourehut use git-send-email[8] to accept and review patches. In fact, basically everything on Sourcehut can be done through emails, like discussions and issue handling.

I think that this is **Sourcehut's biggest weakness** in getting people to switch, because the flow is different and often unfamiliar. I don't think that it's *harder* than the PR flow, but it's significant friction. It makes intuitive sense for discussions and issue reporting, though, and I hope people will give it a shot.

It took me some time to get comfortable with the `git send-email` flow, because it's different and it was intimidating at first. Ultimately, though, I'm a **big fan of the email workflow**. This lets me spend *less* time in my browser, which is a big win for someone with attention/distraction issues. It's **easier to stay focused** on submitting, reviewing, and merging patches when I'm doing that in a dedicated email client instead of in my browser, one click away from dopamine.

The submission workflow is probably not a big deal for the projects I have on Sourcehut. I've gotten a few issue reports on my repos of the years,

---

[8]https://git-send-email.io/

and a sprinkling of contributions, but they're far from having active communities. I do hope that the chess database[9] I'm working on will be able to grow a community, if it ends up being useful, and I'm looking forward to growing that on Sourcehut.

Ultimately, though, I don't see hosting on Sourcehut as being an impediment to contribution. There are **many ways to contribute** and you don't even *have* to use `git send-email` if you don't want. You can email in the patch, of course. Or you can mirror the code to another forge and say "please pull it from here". In either case, **contributors don't need yet another account** to make a contribution. This was one of my factors in switching, because I don't want to force people to create more accounts on more platforms.

21.3 Why not *X*?

There are a few other major forges. It's worth talking about why I *didn't* pick those. This just deserves a few bullet points.

GitLab: - it's super slow in my experience - the features always felt half-baked and there are a *lot* of features - it's very similar to GitHub and I wanted a change - contributors have to make yet another account (or OAuth with a GitHub account, I guess? meh.)

Gitea: - reasonably fast, actually! seems good on that front - doesn't have build support for CI that I could find - feels like GitHub Lite - contributors have to make yet another account (or OAuth with a GitHub account, I guess? meh.)

21.4 When I'll still use GitHub

I'm going to keep my account on GitHub, and I'll still use it for three reasons:

My workplace uses GitHub. I like my job, and we use GitHub, so I will of course keep using it there. There's no reason to push for a change. It's a perfectly fine platform for our needs.

To make contributions to projects on GitHub. My projects are going to be hosted elsewhere but sometimes I'll contribute to projects on GitHub! Maybe bugfixes for things I run into at work. Maybe contributions while pairing at Recurse Center. But occasions will arise when I

---

[9]https://sr.ht/~ntietz/isabella-db/

make contributions to projects on GitHub, and that's going to use my GitHub account.

I have a few projects I'm not moving. Old projects that aren't active, I'm not really investing the time or energy in moving. If these ever have activity, then I'll of course use my GitHub account to manage them.

21.5 Should you leave GitHub?

If this post resonated with you and you're thinking about leaving GitHub behind, you might want to! You can always diversify your forge use without committing 100% to leaving one behind and using the other. That's what I'm doing.

But there are also some very legitimate reasons to stay on GitHub as your primary forge. If you don't have philosophical objections to staying, it's the place to be. For worse, it's the place that employers expect to find activity from candidates. If you're looking for a job, an active GitHub profile can help with that. (If you're responsible for hiring decisions: Please let's talk about changing this!)

I think more people diversifying their forge use would be good for the industry long term. GitHub controls much of the open-source world and they also control much of the software industry. Regardless of their current actions, **this is a major risk** in the long term.

I don't fault any individual for staying on GitHub. But let's normalize using more forges. If you have the ability to switch, please consider it!

---

22 RC Week 9: Parallels of Proofs and Programs

Originally published on 2022.11.19.

I have three weeks left at Recurse Center. This last week was significantly less productive for me than usual, because I've been pretty fatigued and just recovered from a cold. But I still got some work done that I'm proud of. More than that, I'm excited for the coming three weeks!

This week I was mostly fatigued all week, so I didn't do very much coding. In spite of that, I made some really good progress on IsabellaDB[1] through some pairing sessions! A friend reminded me that a few years ago I was *deeply* skeptical of pair programming (I knew it worked for some people, but I was convinced I was not one of those people). This week cemented what I learned earlier in batch: Pair programming is a highly effective tool for getting work done. It's not an all-the-time thing for me, and it's highly dependent on having the right pair for the right problem, but it's a great time.

Through pairing this week, I was able to finish out both a basic move explorer (show the list of legal moves, click one to make that move) and finish out my sparse bitmap implementation. This lays the groundwork for the more interesting features I am building with IsabellaDB. Next up is **displaying win/loss/draw percents** in an opening tree so you can explore openings. After that, building some **filters** to explore openings for a certain subset of games (played in the last 12 months, etc.). And then after that, I'll generalize it to be a **query engine over all the games** so you can do things like search for sequences of positions (want to see how often the Caro-Kann transposes into a French Defense?) or features of positions/games (want to find all the Botez Gambits[2]?).

When I wasn't feeling up to coding this week, I dove into exploring Coq (a proof assistant) and Idris (a functional language with dependent types) more. Right now, I'm getting a lot of energy from exploring this more mathematical side of programming. I'm not sure it'll be sustained energy, but it's really exciting and fun to explore! In particular, doing theorem proving with Coq is just kind of a fun puzzle game and it's addictive once you get the hang of it and the difficulty is at the right level. If the proofs are too hard, you can't really get going in a flow sort of way. But if they're

---

[1] https://sr.ht/~ntietz/isabella-db/
[2] https://www.chess.com/terms/botez-gambit-chess

just hard enough to be engaging but feasible, then it's so delightful and pulls me in.

These two activities—systems programming and theorem proving—came together in a very nice way this week. Last week and this week, working through proofs, there were a few occasions where proofs got pretty difficult. To get through them, there are two general techniques I've been using: - **Break the problem into subparts recursively.** For a proof, this typically is one of a few things. For a particular statement, you may break it down into its cases (a boolean can be true or false, so consider each of those independently). And for a longer proof, you can find an intermediate lemma which you can prove to make the later work easier. - **Update definitions to support your proofs.** Sometimes, a definition is wrong, and clearly needs to be reworked in order for a proof to be possible. I ran into this where I had an edge case that didn't matter until the Final Boss proof; when I fixed my definition, the proof was possible, where before it was not. In other cases, there are equivalent definitions where one will make the proof significantly easier. Usually this lets you avoid intermediate lemmas, and if the proof requires fewer steps from end-to-end it's usually easier to get from the start to the finish, so it makes it a lot easier!

Both of these techniques came into play when I was working on my sparse bitmap implementation, as well.

The first thing I realized was that the way I defined it was not ideal for combining multiple bitmaps. The definition worked and felt elegant, but it was very awkward and hard to reason about when iterating over two bitmaps in parallel. In a pairing session, we changed the definition to an equivalent (but simpler) implementation. This required changing most of the methods implemented on the bitmaps, too, since they rely on the underlying details. But at the end of the day, it was worth it: The implementation of the bitwise operators became so much **easier to reason about and therefore more likely to be correct**.

Recursively breaking down problems also came into play with the bitmaps. This is a common technique in programming in general, so what I'm talking about here isn't shocking. The surprising thing to me, though, was how much **writing my program felt like writing my proofs**. I think it's because it gives me a sense of formalism about how to reason about my code and a mental structure to it. At any rate, exploring proof assistants has made writing programs much easier. That's a win.

There's a strong parallel between the activities of writing proofs and of writing programs. The Curry-Howard correspondence[3] tells us that programs and proofs are directly related. I don't understand the details of

---

[3]https://en.wikipedia.org/wiki/Curry%E2%80%93Howard_correspondence

that yet, but will work toward that through our exploration of Coq. What I do know right now is these activities **are extremely similar** in how I think through things and how I approach them.

Another Recurser presented yesterday on how doing studies[4] (in the art sense where you work through some small pieces in isolation before doing the broader composition) can be a highly effective technique for programming as well. This makes a lot of sense and is a technique I want to try out more deliberately in the future. In a sense, I think I'm already doing it. What is this, if not breaking problems down recursively? (There's a small difference of the study being something you don't intend to reuse directly.) Is there an art equivalent of updating your definitions to support the proof?

It's sort of fascinating the parallels between fields that I think of as typically unrelated. Sure, proofs and programs, we've been exposed to that before. But I'm a little bit mind-blown that there's also a parallel between *art* and programming in form of techniques. This makes me excited to explore other domains and learn how people in other domains work!

See you next week! It'll be a short one, because of Thanksgiving.

---

[4] https://en.wikipedia.org/wiki/Study_%28art%29

23 Measuring the overhead of HashMaps in Rust

Originally published on 2022.11.22.

While working on a project[1] where I was putting a lot of data into a HashMap, I started to notice my hashmaps were taking up a lot of RAM. I mean, a *lot* of RAM. I did a back of the napkin calculation for what the minimum memory usage should be, and I was getting **more than twice what I expected** in resident memory.

I'm aware that HashMaps trade off space for time. By using more space, we're able to make inserts and retrievals much more efficient. But how *much* space do they trade off for that time?

I didn't have an answer to that question, so I decided to measure and find out. If you **just want to know the answer, skip to the last section**; you'll know you're there when you see charts. Also, all the supporting code[2] and data[3] is available if you want to do your own analysis.

23.1 Allocators in Rust

Rust takes care of a lot of memory management for you. In most cases, you don't need to think about the allocation behavior: Things are created when you ask for them, and they're dropped when you stop using them. The times when you *do* have to think about it, the borrow checker will usually make that clear to you.

Sometimes, though, you get into situations where memory allocation behavior matters a lot more for your system. This can be the case if you're very memory constrained (as I was) or if you are trying to avoid the cost of memory allocation. In these situations, Rust lets you define your own allocator with the behavior you want!

The System[4] allocator is the default allocator used by Rust programs if you don't do anything special. It uses the default allocator provided by

---

[1] https://sr.ht/~ntietz/isabella-db/

[2] https://git.sr.ht/~ntietz/rust-hashmap-overhead

[3] https://docs.google.com/spreadsheets/d/1jWv3nzQwncXy0xK_MKmcUkfIT8sbQa02skNxP609az4/edit?usp=sharing

[4] https://doc.rust-lang.org/std/alloc/struct.System.html

your operating system, so it's using `malloc` under the hood on Linux systems.

Another one I've seen referenced a lot is tikv-jemallocator[5], which provides a different `malloc` implementation with some characteristics like avoiding fragmentation. It comes from FreeBSD. I didn't explore using this one other than idly trying it in my main project, where it didn't make any discernible difference in memory overhead[6].

There are some other fun allocators, too, and you can do some really neat things with them. Here are two that I thought were neat:

- bumpalo[7] has a cute name and is a bump allocator that can allocate super quickly, but generally cannot deallocate individual objects; niche in use
- wee-alloc[8] is also cutely (and descriptively!) named and is a "simple, correct implementation" of an allocator for WASM targets, so it generates small code for allocations

There are also a few allocators which help you measure overhead. But where's the fun in that??? Let's do it ourselves!

23.2 Writing an allocator to track allocations

It's tricky writing an allocator that does the useful work of allocation, and there's a lot of nuance. It's a lot easier to write an allocator that wraps around an existing one and records measurements! That's what we're doing.

The thing to know is that we need to implement the `GlobalAlloc`[9] trait. It has two methods we have to define: `alloc` and `dealloc`. We will make something very simple which just wraps `System` without doing anything at all besides passing through data to some record functions.

We start with a struct.

```
/// TrackingAllocator records the sum of how many bytes are
↳    allocated
/// and deallocated for later analysis.
struct TrackingAllocator;
```

---

[5] https://crates.io/crates/tikv-jemallocator
[6] I tried this when someone suggested my high resident memory might be from fragmentation, since jemalloc is better at avoiding fragments. This was before I realized the extent of the overhead of HashMaps, but it did lead me down this allocator journey.
[7] https://crates.io/crates/bumpalo
[8] https://crates.io/crates/wee_alloc
[9] https://doc.rust-lang.org/std/alloc/trait.GlobalAlloc.html

Note that our struct doesn't have any fields. We can't put anything dynamic in it. We'll need some atomic ints and such to track allocations. Since we don't expect to have multiple of these at once, we'll just put those as statics in the module scope. We could put the fields we want in the struct, but it makes constructing it a little more annoying and we won't have multiple allocators at once, so we're just going to make those statics.

```
static ALLOC: AtomicUsize = AtomicUsize::new(0);
static DEALLOC: AtomicUsize = AtomicUsize::new(0);
```

And now we define `alloc` and `dealloc` so that `TrackingAllocator` is `GlobalAlloc`. Implementing `GlobalAlloc` requires marking things `unsafe`. What we're doing here isn't really unsafe, but we satisfy the interface. All we're doing is passing through to `System` and recording it with some helper functions we'll define later.

```
unsafe impl GlobalAlloc for TrackingAllocator {
    unsafe fn alloc(&self, layout: Layout) -> *mut u8 {
        let p = System.alloc(layout);
        record_alloc(layout);
        p
    }

    unsafe fn dealloc(&self, ptr: *mut u8, layout: Layout) {
        record_dealloc(layout);
        System.dealloc(ptr, layout);
    }
}
```

Now we also have to define the helper methods to record allocations. They're as straightforward as can be, just doing a `fetch_add` to record the size of the allocated or deallocated memory into its corresponding counter.

```
pub fn record_alloc(layout: Layout) {
    ALLOC.fetch_add(layout.size(), SeqCst);
}

pub fn record_dealloc(layout: Layout) {
    DEALLOC.fetch_add(layout.size(), SeqCst);
}
```

Now the functionality for the allocator itself is basically in place, and we can move on to using it!

23.3 Using our allocator

There are two things we need to do to use our allocator: Set it up as the global allocator, and add a little bit of functionality to get *useful* data out.

Let's make it the global allocator first. This is the easy bit. Somewhere in your program (such as in main.rs), you create an instance and mark it as the global allocator:

```
#[global_allocator]
static ALLOC: TrackingAllocator = TrackingAllocator;
```

And now that's done. That's all you need to do to change the global allocator! You can see also why we made initialization as easy as possible.

Now to address the ergonomics of use. As it stands, *every* allocation and deallocation will get recorded. That's not quite what we want. We want to isolate certain pieces of the program to measure their allocation separately from test setup and teardown. We also want to record stats from multiple separate runs and report them nicely.

The first thing to do is define a struct for the stats we want to return. We want the total allocation and deallocation, and it would also be convenient to have their difference. This can be calculated later, but let's just include it in the struct for now.

```
pub struct Stats {
    pub alloc: usize,
    pub dealloc: usize,
    pub diff: isize,
}
```

And now we need some helper methods to reset the counters, and get our stats out.

```
pub fn reset() {
    ALLOC.store(0, SeqCst);
    DEALLOC.store(0, SeqCst);
}
```

```
pub fn stats() -> Stats {
    let alloc: usize = ALLOC.load(SeqCst);
    let dealloc: usize = DEALLOC.load(SeqCst);
```

```
    let diff = (alloc as isize) - (dealloc as isize);

    Stats {
        alloc,
        dealloc,
        diff,
    }
}
```

And we have the pieces we need to use this nicely! We can call `reset()` to clear the values before an experiment, and call `stats()` to get them afterwards.

23.4 Putting together the pieces

Let's put together the pieces now and measure the overhead of `HashMaps`! As a bonus, we'll also measure the overhead of `BTreeMaps`.

First let's define a helper function that lets us measure and report on the allocations from a test scenario. This function should take in another function, which will return some data (this is important so that the data *isn't dropped* until after the measurement is complete, or the diff will be inaccurately low). The job of this function is to reset the allocator, run the function, report the stats, then drop the data.

```
pub fn run_and_track<T>(name: &str, size: usize, f: impl FnOnce() ->
↪   T) {
    alloc::reset();

    let t = f();

    let Stats {
        alloc,
        dealloc,
        diff,
    } = alloc::stats();
    println!("{name},{size},{alloc},{dealloc},{diff}");

    drop(t);
}
```

For simplicity we're just printing the results to `stdout`, and the CSV header will be defined elsewhere.

Now let's define our scenarios. For this, we'll first assume that we have constructed some data:

```
let pairs = generate_keys_values(1_000_000);
```

There's a helper function that fills a `Vec` with as many key/value pairs as we want. Each is a pair of a random `u64` (key) and a 100-byte random `u8` blob (value). The particular data here shouldn't matter too much, but I picked something of about 100 bytes to match the domain I originally saw this in.

We'll also have a list of sizes for the tests; later, we can just assume we have a `usize` called `size` which we can use. You can see the full details in the complete listing[10].

Now let's define the baseline. The baseline here is two `Vec`s, one of the keys and one of the values, constructed with *exactly* the capacity we need and no more.

```
run_and_track("vec-pair", size, || {
    let mut k: Vec<u64> = Vec::with_capacity(size);
    let mut v: Vec<DummyData> = Vec::with_capacity(size);

    for (key, val) in &pairs[..size] {
        k.push(*key);
        v.push(*val);
    }

    (k, v)
});
```

And now we can also define our BTree and HashMap scenarios.

```
run_and_track("hashmap", size, || {
    let mut m = HashMap::<u64, DummyData>::new();

    for (key, val) in &pairs[..size] {
        m.insert(*key, *val);
    }

    m
});

run_and_track("btreemap", size, || {
```

---

[10]https://git.sr.ht/~ntietz/rust-hashmap-overhead/tree/main/item/src/main.rs

```
let mut m = BTreeMap::<u64, DummyData>::new();

for (key, val) in &pairs[..size] {
    m.insert(*key, *val);
}

    m
});
```

When we run these (with some additional glue code), we'll get a CSV as output which we can then load into a spreadsheet and analyze.

23.5 The results (I brought charts)

The results surprised me, because I (naively, perhaps) expected the HashMap to maintain fairly constant, fairly low overhead. I was aware that hashmaps in general have a "load factor", but I didn't fully understand how it was utilized. It is used to define when the HashMap will resize to contain more elements. If your load factor is 1, then it will reallocate when the map is full. I think the load factor for Rust's HashMap is something like 7/8. This means that when it has 12.5% capacity remaining, it will reallocate (and probably double, so that the amortized cost of reallocating is O(1)).

If we do some analysis, we can reach a better estimate than my naive unthinking estimate that it would have 12.5% overhead. In fact, it's much higher than that. If the HashMap doubles its capacity when it hits 12.5% remaining (14% overhead), then after doubling it will have 56% free capacity, and the overhead of the extra space is about 125% of the used space. On average, we expect the overhead to be somewhere between those, perhaps around 70%.

How does this compare to what we see in this test?

First we can see the growth behavior of both containers against the baseline:

HashMap and BTreeMap memory usage against baseline

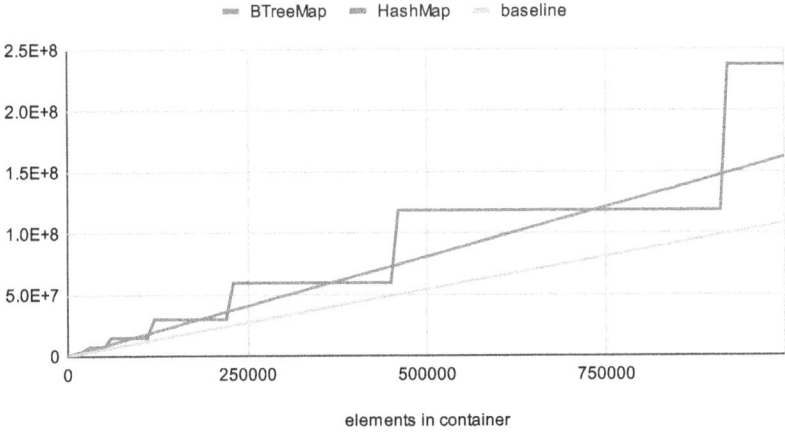

Figure 23.1: Chart of growth in memory usage of HashMaps and BTreeMaps against a baseline

Here we can see that BTreeMaps grow smoothly linearly with the size of the data, while HashMaps are growing stepwise. Additionally, it looks like HashMaps are almost always using more memory than BTreeMaps.

We can see the trends more clearly if instead of the direct memory usage, we plot the *overhead*: as a ratio, how much additional memory is it using compared to the baseline? For the baseline, the answer is 0. From our analysis, we expect the hashmap to average about 0.7.

BTreeMap vs HashMap overhead

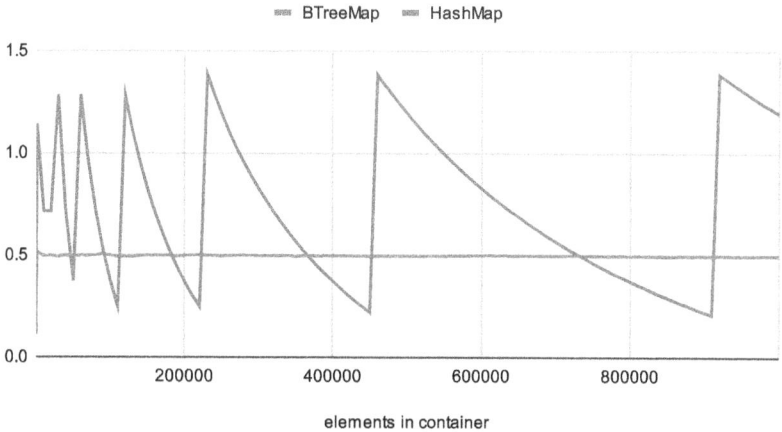

Figure 23.2: Chart of the overhead ratio of HashMaps and BTreeMaps against a baseline

And here we see the behavior more clearly. BTreeMaps do indeed have fairly consistent overhead. On the other hand, HashMaps' overhead swings wildly. It goes up over 1.25 (about what we hypothesized), and drops as low as about 0.125 (also what we hypothesized).

And if we average it? **0.73**. The hypothesis was bang on.

So in general, you can expect to allocate **nearly twice as much memory as your elements alone** if you put them in a Rust HashMap, and about **50% extra memory** if you put them in a BTreeMap.

Hashmaps make a clear space-for-time tradeoff, and it's easier to make that tradeoff effectively if you know how *much* of each you're trading off! Measuring the time tradeoff is left as an exercise for the reader ☺.

24 RC Week 10: Thankful for Family, Missing my Family

Originally published on 2022.11.25.

As I write this, I'm sitting, surrounded by family, recovering from a cold. I wasn't sure what I'd write this week for the RC week 10 recap, since it's a short week. This week I didn't get a whole *lot* of coding done, so it's time for the trope: the Thanksgiving post.

Of course, I'm thankful for my family who I'm surrounded by at this holiday. My two kids bring such joy and love (and frustration and sickness and expenses) into our lives, and I'm so glad they're part of our lives. My wife, who does so much to keep our family rolling and keep the kids fed, clothed, safe, and entertained. My mom and dad, who always open their home to us and help both my wife and I get a break from childcare when we come visit.

I'm also sad about those who I can't be with this holiday.

My grandma passed away a couple of years ago (recently enough that it's both fresh and distant). We honored her this Thanksgiving by making one of our family recipes that she'd make so often for us, called compres galuska[1]. One time years ago, my mom had my grandma teach us how to make it and recorded a video of the process. A few years ago, before she passed but after she wasn't able to make it anymore, I made a recipe with precise quantities from the video and vague instructions she had given us. Our plan this year was to make compres galuska to honor her and to share with my grandpa.

My daughter brought home a nasty cold from preschool (it was mild for her, but not for me). I caught the cold, and so did our toddler son. We brought it with us to Ohio when visiting my family, and consequently... We were not able to share the meal with my grandpa </3. We were also not able to visit my other grandma </3.

---

[1] Compres galuska is a dish of seasoned pork tenderloin, "kinda mashed" potatoes, and dumplings. It's the world's ideal comfort food, in my opinion, which is not at *all* biased by years of my grandma making it for me[^2-week-10]. The name is, we think, of Hungarian origin, but has been changed and we don't know entirely where it came from. It was used by my great-grandparents, and presumably earlier as well.

It's a very weird and unpleasant cocktail of feelings: Being thankful for being around my family and sharing food with them, but also knowing that our being here precludes others from being here.

The cooking worked out well. I was able to spend some really nice time with my mom and dad in the kitchen, cooking this deeply meaningful dish. We were all able to enjoy the fruits of our labor together (my parents, my wife, my kids, and me).

I also missed some of the other family members who can't be here with us. It's easy for my wife and me to attend all of my family's holidays, since her family's holidays are an almost entirely disjoint set from mine. My brother and his wife are not so fortunate, and so they have this tension for every holiday of where they're going to go. It's difficult logistically for them (their jobs are also less flexible than ours), and I feel for them. And I also miss them.

Thank goodness that even though we're far apart, we can get *some* semblance of together time. FaceTime is an absolute blessing at times like this. We're able to video call my grandparents who can't visit us, and we can still get some little dose of togetherness. We can see them, they can see our kids running around (their great-grandchildren!), and it's a great little wholesome moment. And we can keep group messages going, sharing so many little moments.

I've managed to stay offline more this week than usual, but not entirely intentionally—this cold has been kicking my butt. I'm going to go be with my family and spend a little more of this precious holiday weekend with them. And I'm going to re-up my cough drops.

---

25 Tech systems amplify variety and that's a problem

Originally published on 2022.12.01.

I recently read "Designing Freedom" by Stafford Beer. It has me thinking a lot about the systems we have in place and something clicked for why they feel so wrong despite being so prevalent. I'm not sure what any solutions look like yet, but outlining a problem is the first step, so let's go.

25.1 Systems background

First, some background. What's a system? And what's variety?

A **system** is **a group of components and their interactions**. Systems are often used as models for the real world, allowing us to pick out the most important elements and interactions.

Everything you interact with is part of a system and can be modeled as such. *A company is a system.* You can model it with employees as the components, or you can model it with departments as components. *The economy is also a system!* You can model it with consumers and companies and the government as various components which interact.

The **variety of a system** is number of possible states of a system, or of one of its components. Consider monitoring your home's temperature. If you have one temperature probe at your thermostat, that's lower variety than if you have a temperature probe in each room to measure them independently. Similarly, if you have a probe in each room but you average them, the aggregate measure has lower variety than the raw data.

With variety in a system, interactions can **amplify** (increase) or **attenuate** (decrease) that variety. We saw one example of attenuation: Taking an aggregate of some measurements *attenuates* the variety. Likewise, if you decide to add more probes in each room which are not aggregated, that would *amplify* variety.

25.2 Lots of tech amplifies variety

Technology can do either job: it can attenuate or amplify variety for us. The thing is that so *much* of our technology today amplifies variety. Let's look at it through a few examples.

Global news amplifies variety. It used to be that we could see just what's happening in our local town through a newspaper delivered (maybe) once daily. Now we get a firehose of news from all over the world at a moment's notice. Local news once daily was pretty low variety, and instant global news is almost unfathomably high variety.

Social media amplifies variety. It's pretty clear that social media amplifies variety in much the same way as global news. Most social media apps are structured in a way where you can follow basically any person with a public feed. You can follow any famous person you'd like, and you're encouraged to. Instead of having a local view of your local friends, you get to see this gigantic stream of information from all over the world.

Okay, media amplifies variety. What about our other technology? Let's look at tools we use for work.

Chat apps like Slack amplify variety. Slack encourages you to join a *lot* of channels, so instead of a small drip of information you get a firehose of it. And with most cultures encouraging open channels by default, there is a lot of information to take in.

Tools like GitHub Copilot or GPT-3-based writing assistants amplify variety. We're seeing an explosion of tools which use GPT-3 to help you write code or write prose. This amplifies variety in two ways. It increases the variety while you're *using* the tool, because it usually readily shows you suggestions, so you've increased the state of the system while you're using it. It also increases the variety of the system that you're using the tool to contribute into. If you use a GPT-3-based tool to produce documents more quickly at work, that actively increases the variety of the corpus of documents. You write more docs more quickly, so that adds to the firehose again.

Amplifying variety is one side of the coin. It's **not inherently good or bad**, just a property of the system. But variety in the wrong places in a system *can* be bad, and can lead to undesirable outcomes. These outcomes can range from mundane (getting overwhelmed by sensor readings) to catastrophic (total collapse of a company).

25.3 Tech-amplified variety is causing problems

In this case, I think that the our tech-amplified variety *is* causing tangible problems today. This amplified variety definitely has a lot of good points. It's *great* that we have cultures where there's more openness in companies and you can see whatever information you want to see. It's *great* that we're lowering the barriers to writing code and prose and making those easier. It's *great* that we can connect with people from all over the world.

It comes with a cost, and it comes with an opportunity. Every problem begs a solution.

The cost of this amplified variety is that we're pushed beyond human limits. The human brain is finite. There is only so much information we can process, only so much we can take in.

Our brains aren't designed to take in these massive firehoses of information.

Here are some of the problems that I experience and see within these high variety systems, along with how I mitigate some of the problems as they affect my daily life:

- *Global news induces anxiety and depression.* We see all the problems of the world on broadcast, and we see the good news from only our immediate circles. This imbalance contributes to mental health crises.

  Mitigation: stop consuming global news on a regular basis. We're taught that it's important to consume global news to "stay informed", but practically speaking, there's nothing actionable I can do with this information anyway. I have to disconnect to preserve my mental health. I'm still on the lookout for a way to get news summaries, in context, on a slower cadence.

- *Social media distracts from other tasks.* When I was on social media (my vice was Twitter), it was what I turn to for a brief distraction. With high variety, there's *always* something new and interesting on it.

  Mitigation: get off social media. For me, the mitigation is to go cold turkey off of it. I'd like to find a balance here, and I hope that with the rise of Mastodon, we may see humane social media which isn't a dopamine factory by design.

- *Too many documents, emails, and chat threads to read at work.* There's simply too much information produced in even a small company for me to process all of it.

Mitigation: read a limited subset, and read anything that someone specifically calls my attention to. With limited time, I pay attention to a few "blessed" channels that are highly relevant to my daily work. For the rest, I sometimes skim but usually let it go by unless someone calls my attention to something, which I then go engage with.

One common element of all of my mitigations: They *attenuate the variety* of the system that's my life.

The problems of high variety come from an **imbalance of varieties**. The variety my brain needs is much lower than the variety offered by global news, social media, etc. so I have to attenuate those varieties to bring them back to something I can deal with. This is true for systems in general. If a component is experiencing higher variety than it can handle, it's going to experience negative effects. Attenuating variety for that component is the answer. Similarly, if a component *can* tolerate high variety, you have a bit of waste if you feed it only low variety; it could do so much more.

Big problems come in when variety is left unattenuated and is higher variety than the component it's fed into. I think this is why we see such discord and division in the US, contributed to by social media. But I don't know for sure.

25.4 What led us here

Spoiler: it's **incentives**, it's always incentives. But also, **attenuating variety is hard**.

We got into this mess because the incentives of our (capitalist) economy lead to prioritizing amplifying variety. And when we try to attenuate it, that's a harder problem, so we can't do it as effectively—especially without resources, because those resource are poured into amplifying variety.

The main incentive in a capitalist economy is making a profit. Right now, the main way that's done through tech is through monetizing your **attention**.

Anything that's funded through advertising has a clear model: Get you to spend more time in their app, and they make more money.

Other monetization strategies without advertising often end up grabbing attention anyway, though. GitHub has all the dopamine hits on it to *get you to use it more*, which makes their platform more valuable. They can then use that platform to get enterprise sales and other paid features.

They also use that platform's wealth of data to create new products, like GitHub Copilot.

Even products where you pay directly want to keep your attention. Why do brands push their messaging so much into your inbox? To keep your attention so that when you decide to spend money, you spend it with them. And products you're subscribed to, like Netflix, want you to actively use them as much as possible so that you keep paying for them and feel like you got your money's worth.

So, these products are amplifying variety. If it's so nice having attenuated variety, though, why don't we do that? Some consumers would surely pay for that.

The problem is: It's brutally hard. Let's look at the news as one example. If we collect all the news worldwide, that amplifies variety. Now we want to attenuate it to make it consumable without problems. A few ideas for how to do that are:

- *Have a team of writers/editors condense the news down into something intelligible.*

  This is likely very expensive (on top of collecting the news, you must pay again to condense it). It can also create the perception of the problem of bias: What gets included and what does not? And any summary will have some human perspective applied for what's important. Doing this with AI isn't a great idea, in my opinion, but that would also create problems with bias.

- *Filter to subsets based on relevance/interest.*

  News posts could be tagged with their topics. This is probably already done, as newspapers are organized into sections, and you could also use a topic model to help generate these automatically. This contributes to the problem of filter bubbles, though. It defeats one of the general benefits of being a news-consumer, which is to get broad exposure to more topics. Additionally, it only attenuates in one way by exposing you to fewer topics, but it keeps the variety very high within those topics. So it's an incomplete solution.

In general, I believe it's much easier to create a system that amplifies variety than to create a system that attenuates the newly-created variety. Any form of attenuation is breeding ground for novel *new* problems, and it's just expensive and hard. Given the difficulty and the incentives at play, is it any wonder that we keep making systems steal more of our attention and amplify variety?

(No, dear reader. It is not a surprise.)

25.5 Where do we go next?

Okay, so what do we do?

I don't know. Like I said, it's a *hard* problem. It's certainly not going to be solved in one schmuck's blog post.

One thing I do know, though, is that we can fight back and we can change the system. Part of how I mitigate the attention-stealing techniques of apps that amplify variety is by **practicing mindfulness** and being aware of where my attention goes. With that awareness, you can choose to prioritize products (like Sourcehut) which respect your attention over products (like GitHub) which try to take it. You can also prioritize using products like GitHub how *you* want instead of how the designer wants you to, although that's very hard. They're putting a lot of money into changing your behavior.

The one thing I do know is that by talking about the concepts here and the problems, we can increase awareness. Maybe we can shift how systems are designed. Maybe we can shift how they're used. But we can certainly talk about it, build awareness, and try to *collectively* come up with solutions.

I think part of the long-term solution is alternative incentives. We see this happening with structures like benefit corporations[1], which prioritize other incentives in addition to profits. We're seeing a broad global shift toward more focus on sustainability, because consumers are demanding it. We can demand more respect for our attention, and shift the system.

Let's design, build, and buy humane systems which work for us rather than exploiting us.

---

[1]https://en.wikipedia.org/wiki/Benefit_corporation

26 RC Week 11: Learning is best when multiplayer

Originally published on 2022.12.03.

As I come up on the end of my batch at Recurse Center[1], I've been doing some reflecting on my time here. One of the standout themes is how much I've learned through struggling *with* other people. In particular, this learning together has make some difficult topics[2] approachable, where I may have given up or gotten stuck on my own.

This week, we were working through a couple of chapters of Logical Foundations[3], a book which teaches Coq and its related concepts. The earlier chapters were for the most part smooth. I could probably have gotten through them on my own[4]. But chapter 5 (and to some extent, 4) was where we hit an absolute wall.

Some of the proofs in chapter 5 were just absolute beasts to get through until we figured out the particular techniques we needed. In particular, we had to remember to always include `eqn:E` (or similar) for every `destruct` tactic; it doesn't hurt (just adds more into the context, which can be overwhelming), but if you *don't* do this you sometimes get into a situation where you lack what you need in the context, so the goal is not provable! Getting to this technique required a lot of back and forth between a couple of us.

I think there are a few things going on which make learning so much more effective with a peer group:

- **You have someone else to explain things to.** Just by trying to explain something, your own understanding will get better[5]. I first realized the power of this when I was a math tutor and found myself getting *better at math* by explaining material to other people. It shored up my knowledge of the foundational material, and also

---

[1] https://recurse.com
[2] https://en.wikipedia.org/wiki/Coq
[3] https://softwarefoundations.cis.upenn.edu/current/lf-current/index.html
[4] Even if possible, it would have been less fun and less effective. Reviewing the chapters with others has always helped me enhance my understanding (by explaining) and learn new things that I was missing (from other people pointing out things).
[5] This is a large part of why I write on this blog, too. Writing is thinking (for me, at least), and is a vital part of how I learn and understand.

gave me insights into multiple ways to explain things, which aids understanding.

- **Talking through problems helps you get unstuck.** Sometimes, your learning partners will see the problem and be able to nudge you in the right direction! Even if your partner doesn't have a solution, though, you can get unstuck just from talking. This is like **rubber-duck debugging**, where by saying something out loud you often get insights into the solution.

- **You see other approaches.** There is rarely only one right way to do things. By working through problems with other people, you get to see multiple approaches and get a richer understanding of the problem and solutions.

- **You have accountability.** This one is big for me. If you know that on Thursday, you and your "axiom amigos" are going to meet to discuss the chapter, it lights a fire to actually get through the reading and the problems. When doing things on your own, it's a bit harder to keep momentum.

This doesn't work for everything. Sometimes I'm going to have to just chug through material on my own. But I can get a lot of these benefits without having a formal group that's going through the same material:

- I can write on my blog to explain things to other people
- I can talk through problems with other Recursers and friends when I get stuck
- I can read other people's blog posts or texts to see other approaches

Accountability is the big one that's lacking when learning entirely on my own. One way I try to keep that is with a schedule for when I put up new blog posts. This motivates me to learn *something*, so I always keep forward momentum in some direction.

I think this multiplayer learning is one of the best parts of Recurse Center, and one of the hardest things to get outside of it. But I can run book clubs at work[6], join some groups of future RC batches, and keep learning with friends (at a lower intensity) post-batch.

---

[6] ./blog/running-software-book-reading-group/

27 Building Molecule Reader in one day

Originally published on 2022.12.07.

Reading on screens is very difficult for me. I just cannot focus on the articles, especially when there are notifications coming in or even just other content on the screen[1]. I have a reMarkable tablet (RM)[2], which I love dearly[3] and much prefer to read on. But it's annoying getting articles onto it.

To put a blog post onto my RM, I copy the link from Firefox (my usual browser), open Chromium, load the page, and print it with the "Read on reMarkable" printer (which is only for Chrome-based browsers). And when I have five or ten articles I want to read, I have to repeat this for each one manually. Ouch.

It's also annoying how the articles end up. I'd like to have them all tidy in one folder where I can read them, or even in one continuous document. If I send each one individually, they just litter the home screen (since you can't print to a specific folder) and displace other things I'm reading.

I decided to solve this by writing a web app to bundle up my reading and send it to my RM!

27.1 Preparing to build

I knew generally what I wanted: get a bunch of articles, merge them into one PDF, and send that PDF on my RM. And I wanted to do it in Rust. Oh, and only spend one day on it.

That last requirement is the tricky one. The scope could easily get too big, or I could end up cutting critical features for the deadline. Balance was going to be tricky, and this was ambitious. At this point, I was not expecting to finish, but hoping to get *something* working.

---

[1]For some reason, I don't have this problem while coding, which I can hyperfocus on, but I do not get into that same state while reading.

[2]https://remarkable.com/

[3]It's the only electronic device I've used every single day since I got it in 2018. It replaced a handful of paper notebooks I carried everywhere. I now have two, since I got the newer model as well! The original is my book/paper/blog reading/annotating device, and the newer one is for my notes, todo lists, sketches, illustrations, etc.

Since I came up with the idea last week and did the build yesterday, I was able to spend the weekend thinking about requirements and doing some idle searching on helpful resources. I went into the day pretty confident that I could find libraries to do the RSS feed parsing, but I was most unsure about generating the PDF.

My plan going into the build day was to pair program with other people to keep me making progress and start from the PDF generation, since that had the most unknowns. Everything after that was going to be improvised.

Like any good project, I also gave it a good name and made a repository. The project is Molecule Reader, because it bundles up Atom feeds, and what else would you call a bundle of atoms? You can check out the repo[4], with no apologies for the quality of code.

27.2 Building it

Yesterday, I started the day by announcing at my morning RC check-in that this was my plan, and I put out a call for pairing. The first thing I tackled was PDF generation. This turned out to be the easiest part, after I switched up my approach.

My original approach was to use fantoccini[5] (or similar) to use WebDriver to control a browser to render the articles into individual PDFs. I wanted to do this since I knew browsers in print mode would render the articles pretty well, where I was less confident in any reader-mode shenanigans I might be able to pull off in pure Rust.

Instead, I wound up using headless Chromium to render the pages to PDFs! This took me some time to figure out since the command-line options for this were hard for me to find documentation for, but once I got it, it has worked pretty flawlessly. With one little Rust function, I can generate a PDF for the page at a given URL:

```
/// Takes an item and generates a PDF for it. Will return an error
↪  if it fails
/// and an empty result otherwise.
///
/// Assumes that you have `chromium-browser` installed on your
↪  system.
pub fn url_to_pdf(url: &str, output_filename: &str) {
    let print_arg = format!("--print-to-pdf={output_filename}");
    match Command::new("chromium-browser")
```

---

[4]https://sr.ht/~ntietz/molecule-reader/
[5]https://github.com/jonhoo/fantoccini

```
        .args(["--headless", &print_arg,
        ↪  "--virtual-time-budget=10000", url])
        .status()
    {
        Ok(_) => {}
        Err(e) => {
            tracing::error!(error=?e, url=url,
            ↪  output_filename=output_filename, "error while
            ↪  printing page");
        }
    };
}
```

I also found a simple utility for collating the individual PDFs into one. There's this tool called pdfunite which is installed on all my systems already. I run Fedora, but I don't know if this is standard or something that comes with another tool I have installed. At any rate, it was super convenient, and this only has to work on my machine(s)!

```
/// Takes a sequence of filenames and collates them into a combined
↪  PDF with the
/// specified string. Returns an error if it fails and an empty
↪  result otherwise.
///
/// Assumes that you have `pdfunite` installed on your system.
pub fn collate(filenames: &[String], combined_filename: &str) {
    let mut args: Vec<String> = filenames.to_vec();
    args.push(combined_filename.into());
    match Command::new("pdfunite").args(args).status() {
        Ok(_) => {}
        Err(e) => {
            tracing::error!(error=?e,
            ↪  combined_filename=combined_filename, "error while
            ↪  collating");
        }
    }
}
```

With the hard part out of the way by early morning, I was able to move on to the rest of the build. Scraping the RSS feeds was up next, and this was where things were a little hairy. I wanted to process both RSS feeds and Atom feeds, so I used a wrapper library[6] which tries parsing in both formats and gives you a parsed feed in one or the other.

---

[6]https://github.com/rust-syndication/syndication

After that, I built the web application itself! This part was pretty straight-forward with actix-web[7], and the type system made plumbing together the forms a delight. I created three pages, and all the templates are parsed and checked at compile time, so I always know that my page templates will render if the app compiles.

A few folks were around to pair with me for the feed parsing and for the web app build, and it was really nice having company and having ideas to keep me moving forward[8]!

I got something hacked together that worked, but had a lot of sins in it, before 5pm that day. That's a success. I did come back that evening and polished things up, including a refactoring that resolved the lingering performance problems from the original hacky data storage.

And it is pretty nice, in my *totally unbiased* opinion. There's almost no CSS (there's one rule, which strikes through any articles which have been printed; I keep them visible in case I need to reprint), but it works.

---

[7]https://actix.rs/

[8]This was *particularly* helpful while debugging. Huge shoutout to Conner, who helped me debug something and sped me up a lot as a result, and spent a long time pairing with me yesterday.

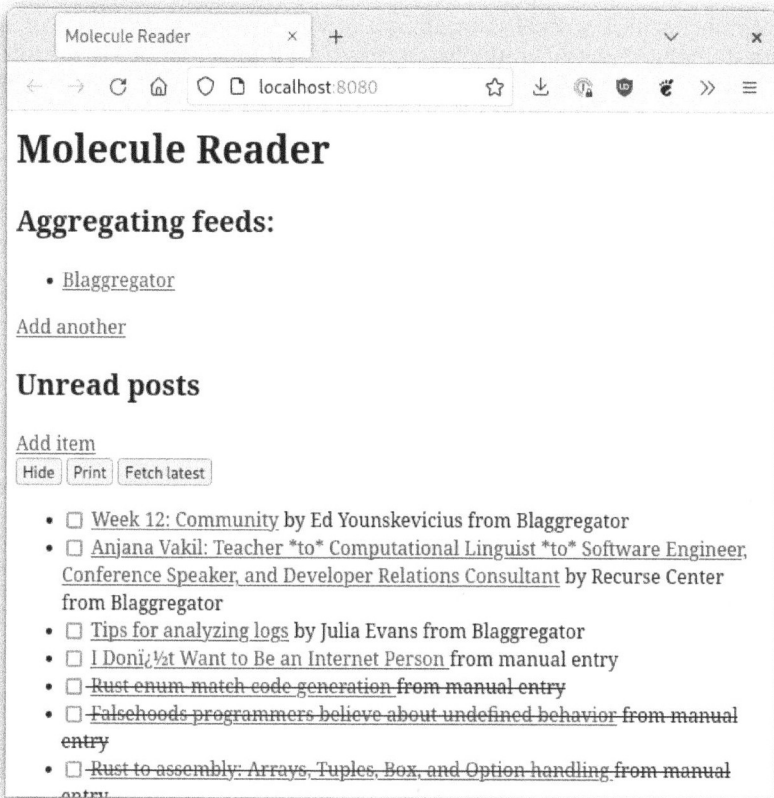

Figure 27.1: A screenshot of Molecule Reader, showing the feeds which are aggregated as well as the latest items.

27.3 Reflections

This was a really fun experience. I'll probably do it again, balancing it against the intensity of the experience. Today I was pretty tired, and I think the one-day build yesterday was a big part of why I'm tired today. (Also shoutout to my kids for why I'm tired today. Love you, and you're a lot of work.)

I took away from this a few things.

Building something useful in one day is possible. I've wanted a small suite of personal web apps for a while to do little things. This has made that possibility much more tangible by reminding me that building something useful quickly is doable. I didn't have to build some of the more complicated pieces (login, permissions), and I don't have to deal with robust error handling since I can always check the logs. **I'm excited to see what I build next!**

I can prototype quickly in Rust. I used to assume I'd better build a prototype in Python, or maybe Go. Those were the languages I can go quickly in, since the type checker isn't along for the ride. That's not the case anymore! Since I've spent the last 12 weeks working in Rust, I think I'm as productive in it as in any other language I know. My prototyping speed in Rust is on par with my other main languages, but with the advantage of the type system making later refactoring much safer.

Rapid prototyping is exhausting. I'll definitely do this again, but I'm pretty tired. There are two aspects of that. One was the intensity of it! Building something useful in one day is a lot of constant hard thinking, and that's draining. The other is that I was pairing for over half the day, and that's pretty draining for me as well. I love pairing, and I can't handle a ton in one go. This draining aspect is something I'll have to keep in mind and maybe stretch things into two-day builds!

I'm proud of what I've built, and excited that it exists. Now please excuse me while I go read 76 pages of blog posts on my RM.

28 RC Week 12: What's Next, and Speedrunning Crafting Interpreters

Originally published on 2022.12.10.

And that's it. My batch at RC ended yesterday. I have so many thoughts and feelings from this time, but it's going to take time to coalesce them all. I'll write up my Return Statement[1] in a week or two, but for now, here's what I was up to the last week!

Mostly, this last week was an attempt to speedrun Crafting Interpreters[2]. This book has been on my shelf for a while, and I got started on it after I decided to stop learning Idris. A friend from this batch has done really cool work going through Crafting Interpreters, so I wanted to see how much I could get through while we can still easily pair program on it.

Turns out, a lot! In the last 1.5 weeks or so, I read through the first 11 chapters and implemented everything from the first 10. All that's left is doing chapter 11 (which should fix a hole in the semantics and improve performance) and then read two chapters focused on classes! It'll be really cool to see how object-oriented programming can be implemented at the language level.

Overall this book has been a great experience so far. So far the benefits have been: - **Greater mechanical sympathy for parsers.** It's easier to understand errors coming out of a parser having written a basic one! Now when parsers leak some details out in errors, it's less confusing. This alone is **a great reason to read the book.** - **Got over my fear of parsers/interpreters.** Before this, parsing was very intimidating. I wrote a little parser for my chess projects to load in PGN files[3], but that was hard and confusing and didn't work well. Now that I've seen a reasonably-structured parser and written it myself, I'm a lot more confident that I can and will write more parsers in the future! I'm currently planning on implementing a query language for my chess database[4]. - **Gaining a better appreciation for the nice things we have.** After

---

[1]Return Statements are a tradition where Recursers write a post about what they did there and some reflections. I'm waiting a few weeks for everything to gel before writing mine, because right now I'm a maelstrom of feelings.

[2]https://craftinginterpreters.com/

[3]https://www.chessprogramming.org/Portable_Game_Notation

[4]https://sr.ht/~ntietz/isabella-db/

writing this much of a language, honestly, I'm extremely impressed and grateful that **other languages work well at all**. This stuff is *hard*.

I'm going to keep running through Crafting Interpreters over the next few weeks, but with less intensity since I'm going back to work on Monday. I think part 2 will be just as fruitful as part 1, since I'll get to see how a (bytecode) compiler works! Maybe my chess database query language will compile down to bytecode for the query engine.

This week also contained a one-day build of a useful tool for my own use. Since I wrote about that earlier this week[5], I won't say much here except that I think the reports of Rust being bad for prototyping are are greatly exaggerated.

This week has also led to me leaning into doing type-driven development with Rust, and leveraging tooling to generate a lot more of my code for me. (Not AI generation, but automatic generation of some boilerplate.) I'll write more about that soon, too.

The rest of this week was coffee chats with folks and reflections on our batches and what is next for us. I'm really excited to see what all my new friends end up doing next. And I hope they stay in touch and stay active on Zulip.

As for me, I spent some time this week making sure that my life is structured in a way that means I can keep doing some of the most rewarding things from RC. Specifically, that means **I've let go of some projects** (Advent of Code, learning theorem provers with friends) to be able to focus on the things that are most important to me. Here's what I'm going to keep on with: - **Consistent writing.** This has been tremendously rewarding, and I'm going to keep up with it post-batch as well as I can. I've setup a dedicated chunk of writing focus time each week, with some folks joining in. I'm optimistic. - **Crafting Interpreters and other technical books.** This book is such a joy, and it's inspirational for me. This is the sort of writing I aspire to eventually. I'm going to keep up with it and then work some other technical books, like CPython Internals[6] (we have a reading group starting in January!). - **The chess database.** This project has taught me so much, and is useful to boot. I'm going to keep going with it as a slow burn so that it's sustainable and keeps going. - **Coffee chats.** Everyone at RC has been so great, so I'm going to keep in touch with folks. A lower intensity and lower frequency, but still there.

In the next two weeks, I should have a Return Statement posted. I have a few other blog posts in the works, too. If you made it this far, thanks for reading!

---

[5] ./blog/one-day-build-molecule-reader/
[6] https://bookshop.org/p/books/cpython-internals-your-guide-to-the-python-3-interpreter-anthony-shaw/16978914

29 Working with Rust in (neo)vim

Originally published on 2022.12.16.

I've been using vim for nearly as long as I've been writing code. My first introduction to it was being thrown in the deep end in 2009 by my Intro to CS lab assistant, who told us to write our programs using vi[1] on the department servers. Why he told us that, I have no idea. But I got used to switching into and out of insert mode, and also how to save and quit.

At my internship in 2011, I learned to use vim in earnest. The project I worked on thrashed system memory by running HBase in a test suite over and over, and my work would routinely crash Eclipse[2] as a result. I don't remember if my mentor suggested it or if I used vim on my own, but he did encourage it. He urged me to learn *proper* vim and disable the arrow keys to get used to navigating with the `hjkl` keys. That got me to learn it quickly through immersion and I fell in love.

Now vim[3] is how I think about text editing, so I'm mired in it. I'm not leaving vim if I can help it, so I've figured out how to use it effectively for the development I'm doing. And these days, that's Rust as often as I can justify it!

I used to use vim in a pretty bare-bones fashion, but I've slowly been layering in more plugins. (Still far fewer than some people I know, but it cannot be described as a minimalist setup.) One of my batchmates at Recurse Center is a vim aficionado and helped me get a really snazzy setup.

All told, I think vim provides an amazing editing experience for Rust (and in general). This is how I develop Rust in vim!

---

[1]Yes, I'm aware that vi and vim are different. I think that vi was symlinked to vim on that system, but I don't know. It doesn't really matter for this story.

[2]IntelliJ was around, but I don't remember people using it. At least I wasn't using NetBeans. I did try. It was worse.

[3]I use "vim" to refer to both vim and neovim. In this article, you can just assume I always mean neovim, since that's what I use exclusively these days.

29.1 Plugins and configuration

First let's look at what plugins are installed. (This is all in my public config repo[4].)

Some general development quality of life ones: - nerdtree[5] for file navigation. - fzf[6] for searching for files by name or content - obsession[7] for saving and resuming sessions more easily - editorconfig[8] to setup spaces/tabs etc. based on the current project

The Rust-specific ones are: - rust-tools[9] to setup the Rust LSP automatically for you - nvim-lspconfig[10] for configuring neovim's LSP (`rust-tools` depends on this one) - nvim-cmp[11], cmp-nvim-lsp[12][13], and cmp-buffer[14] for completions

29.2 Workflow

It's hard to describe a coding workflow through just prose, so I'll use some examples. These are some of the things I run into every day while writing Rust.

The overall workflow is probably familiar to terminal-dwellers, but is different from what IDE-users do. Where an IDE contains all the things (you run your terminal, your tests, your text editor, all in one place!), that's what tmux[15] does for me. When I sit down to code, I start a new tmux session with a window for git commits/logs, another for my editor, and usually another for my tests.

Once I have my editor and test watcher going, the general workflow is: - Write some code in vim, ideally with tests - Check on the build/tests, iterate until it passes - Check clippy[16] for any lints - Write a messy commit message - Repeat until I have a unit I want to merge - Push it my git forge, and squash/merge when CI passes

---

[4]https://git.sr.ht/~ntietz/config/
[5]https://github.com/preservim/nerdtree
[6]https://github.com/junegunn/fzf.vim
[7]https://github.com/tpope/vim-obsession
[8]https://github.com/editorconfig/editorconfig-vim
[9]https://github.com/simrat39/rust-tools.nvim
[10]https://github.com/neovim/nvim-lspconfig
[11]https://github.com/hrsh7th/nvim-cmp
[12]https://github.com/hrsh7th/cmp-nvim-lsp
[13]These names will always trip me up because they have "cmp" and "nvim" in different orders, and somehow that doesn't stay in my head.
[14]https://github.com/hrsh7th/cmp-buffer
[15]https://en.wikipedia.org/wiki/Tmux
[16]https://github.com/rust-lang/rust-clippy

A lot of this workflow is not unique at all to vim, tmux, or any of the other tools—it's just plain software engineering. I think the more interesting things are how I do some specific things while using vim.

Opening a file. The scenario is I know that a file exists with some code I want to modify. If I know the name of the file, I usually use fzf (bound to `control-f`) to search by filename and open it directly. On days when I want to do some sightseeing (more common for codebases I'm not as familiar with, to stumble upon things), I'll navigate through the file tree from nerdtree, but this is rare these days. And in the cases where I don't even know the name of the file, but just something in it, I use ripgrep (bound on `control-g`) to search through the file tree to find any files which have that content! The beautiful preview panes are a big help in finding things easily.

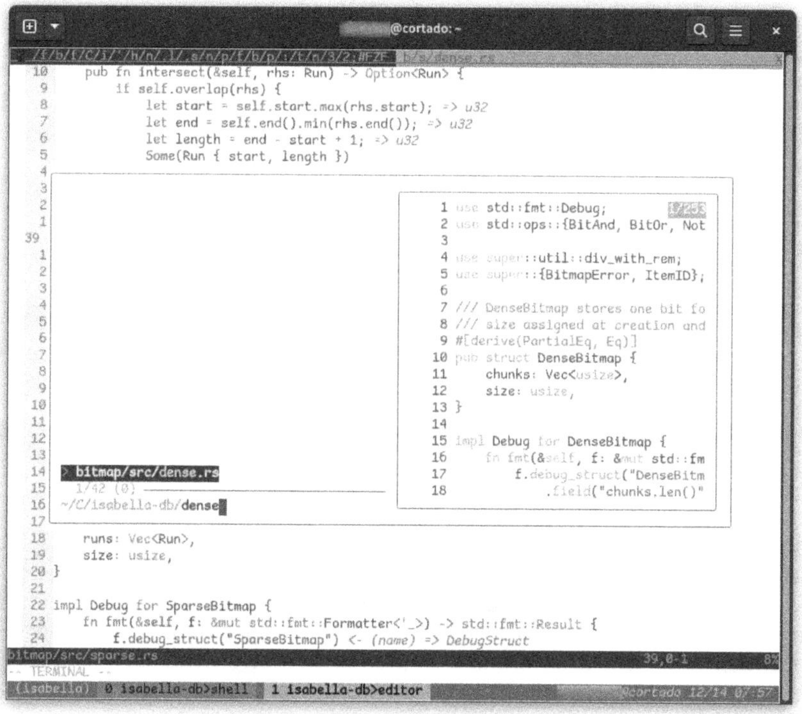

Figure 29.1: Screenshot of the vim text editor showing file search with fzf

Creating a new file. This is where I turn to the trusty friend, nerdtree. (Usually. There are tricks to make this faster with the Rust tooling.) I open up nerdtree, navigate to where I want the new file, and enter a name. This is the same for moving or renaming files.

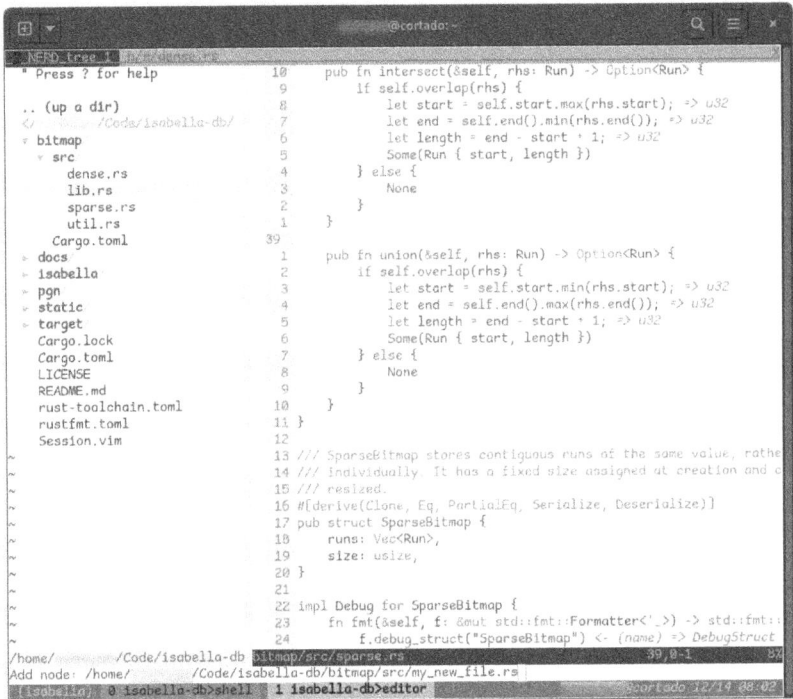

Figure 29.2: Screenshot of the vim text editor showing file operations with nerdtree

Writing code. This one is pretty common, so I won't spend a lot of time on it. I write code in the idiomatic vim way (I think?), and I don't do anything particularly unusual with it. I do avoid certain things (code folding) which I find confuse me more than help me. I just keep it simple as much as I can, and spend complexity points on the *really* valuable things.

Formatting code. This is one where I lean on the Rust tools! I have bound control-f to run the formatter. This is a good balance: It doesn't run automatically (it's jarring when things change out from under me), but it is also so easy to do that I do it often. It's a great part of my workflow! I can write something with odd formatting, then hit control-f and *boom* it's pretty.

29.3 Cool Rust code actions

One of my favorite things now is using code actions (provided by Rust's LSP and the neovim integration). They let me make a lot of common actions faster, and are especially powerful combined with Rust's type system!

Create missing files. From the above section you can probably tell that creating a file was one of my slower manual actions. Searching for files: super fast! Making a new one: manual and slow. This is a little bit easier with code actions. I just refer to the file (usually `use my_new_module;` or something in `lib.rs`), then I press `\a` and a code action is available to create the missing module!

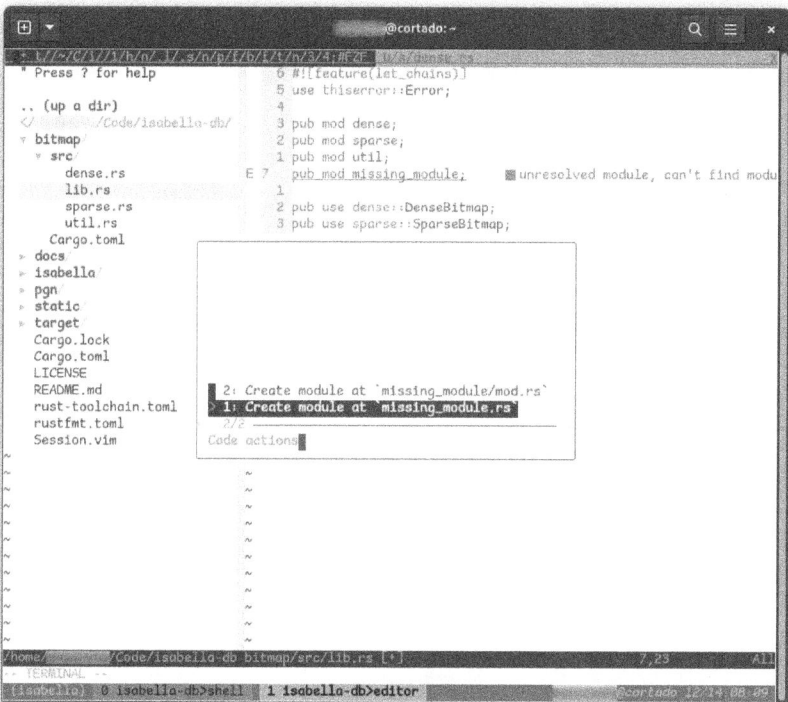

Figure 29.3: Screenshot of the vim text editor showing a code action to generate a missing module file

Generate missing methods. This is similar to the above. My old workflow was often to think about what method I would need and write that (at least a stub with a `todo!()` inside of it) that I would then use in another place. That would get the fewest compiler errors as I went. With code actions, that's flipped on its head: I first write the places where I *use* the method, then I let it generate the missing method. The advantage of working this way is that it can usually write the entire type signature for me, since Rust has a strong type system and there's a lot of information to power its guesses. (If it can't guess correctly, it does something conservative, like leaves a hole for the type.)

Generate required members for a trait impl. Oh yeah, no more looking up the docs to know what I need to impl a trait. I can just make the computer do that work for me. This is really handy for things like `std::fmt::Display` where I might not remember the exact type signature, and even more so for things like `IntoIterator` where there are also types I have to define inside the impl block.

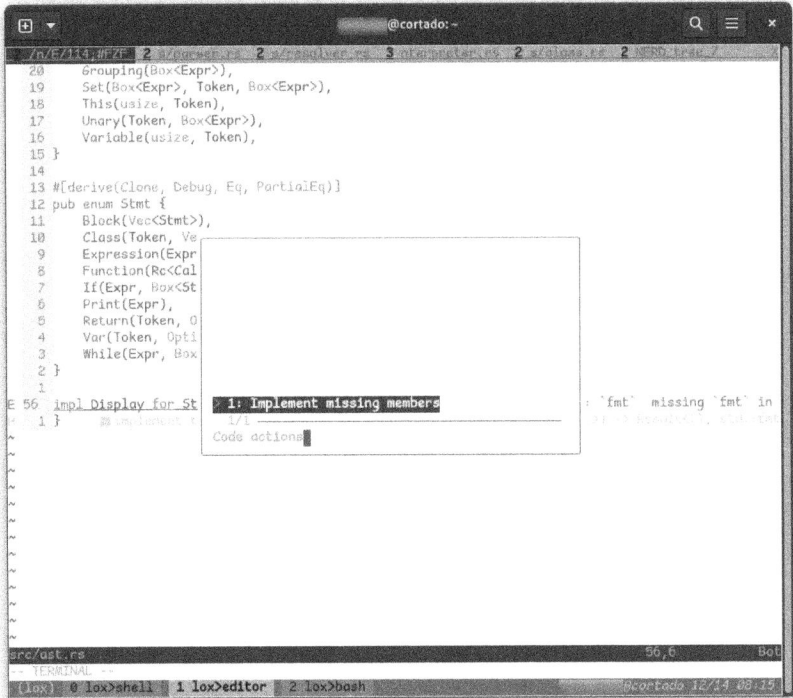

the way I think about my interactions and changed my assumption that interactions with people necessarily drain my energy.

Okay, that's probably it for the *really* squishy personal stuff. Now onto some of the **more programming-related stuff**. I mean, it's still squishy personal stuff.

One of the things I did during RC was **learn in the open**. I wrote a lot on my blog, and I posted check-ins every day where I detailed what I had done, what I was working on and thinking about, and the challenges I was running into. This was very new for me. In the past, I've always waited to talk about details of things until I'm sure that they're "ready." This was an element of being **afraid to fail or be wrong**, and some amount of judgment that may come along with that. In contrast, during RC, I learned to put myself out there and I got to experience how helpful it was—both to me and to other people. I'm going to **keep doing this** as much as I can.

Related, I **learned how to pair program more effectively**. More fundamentally, I learned what my hangups with it were. In my first week, we did some pairing exercises, and I noticed that I would freeze up during pairing. After getting off the pairing call, I'd often have this experience of immediately realizing what was blocking us, or look up one thing and figure it out. But during a pairing call? No chance.

With a lot of introspection and advice from batchmates and the facilitators (shoutout to James Porter in particular!), I was able to figure out what was blocking me from making the most of pairing: It felt *performative* to me. I think this comes from many years of being an honors student or star employee and being *expected* to be right and have the answers, and I internalized that. But that's not helpful, and none of us know all the answers. I learned how to be vulnerable and how to take a beat to think, to read the docs, to take a break. That shifted my relationship with pairing. It's still a *somewhat* draining activity, but it's going to be a regular part of my work going forward. I love it.

I also **learned what I want to do with my time and my life**. I mentioned above that I'm a writer. I used to say I want to start a startup or other business. It felt like the thing I should do in order to "control my destiny" and be able to choose what I work on, to an extent[5]. No, I want to keep doing what I did at RC: **learning, experimenting with technology, and writing about it**. I'm fortunate to have a job where I have a *lot* of flexibility (our leadership is pretty forward-thinking) so I have reasonable hours and four-day workweeks. I'm going to make the most of

---

[5]In reality, I can't think of any worse way to choose what I work on than starting a startup. I'd have to focus on all the other things that go into it rather than the things I really want to work on.

my time and lean into this as much as I can on my Fridays, mornings, and evenings.

Related, I learned how to **engage with projects consistently** and **let go of them when I'm done with them**. I don't mean when they're done. I used to feel this need to ship projects, to completely finish them (as if a thing is ever finished!). Not anymore! Now when I've gotten out of a project what I wanted to learn, and maybe written about it, then it's done and I can let go and move on. I love you, projects, and I have to move on to another one now! It's very liberating, and it lets me learn about so many different things.

Similarly, RC put into focus *how* I spend my time, so I learned to focus on the things I *really want to do* and engage with those projects consistently, not with the ones that aren't how I want to spend my time. I've cut out any exploration of devops things in my personal time because while it's useful, it's not what I want to spend my time on! And I've leaned into learning about programming language interpreters right now, since they're so interesting and fun.

Another one which I've written about before[6], but is worth mentioning again, is my **evolution of thinking on open-source**. Letting go of shipping things as potentially-commercial projects meant I could really lean into copyleft licenses. It's really freeing to think about starting projects and realize that I can make them, license them under a copyleft license, and rest easy knowing that if the code is useful and someone wants to use it, they can. It's a big mindset shift to also chain myself to the mast of open-source and not *allow* myself to make a proprietary version of my code (once there are any other contributors).

I think that's the major formative and life-changing things from my batch. I'm sure I'm forgetting something, though. So *much* happened in the batch.

Which naturally leads into...

30.2 How did I spend my time during batch?

My time was generally split into four buckets: working on my projects, talking to individuals, working with groups, and going to events.

A typical day would generally start at 8am with my check-in call. Then I'd probably have a coffee chat, or take our toddler to preschool. By 9:30am, I'd be at my desk and in the swing of things, so I'd spend the morning making pretty good progress on my project or pairing with someone on it or theirs. Then I'd take lunch, sometimes over a meaty subject

---

[6] ./blog/my-evolution-open-source-licenses/

like proving theorems in Coq[7]. In the afternoon, there were often more groups, and I'd have some coffee chats or pairing sessions and work on my projects. I wrapped up my days at 5pm. I'd usually come back and write my check-ins around 7:30pm, after our kids were in bed.

But days were variable! I bounced to so many different things, and took opportunities to pair or attend events as they arose, so I didn't really stick to a firm schedule. I was also kind to myself and listened to my brain. If I wasn't into something that day, I'd work on something else.

So, here's the laundry list of things I did:

- Wrote a key-value store[8] which can beat Redis's performance multi-core and comes close on single-core.
- Wrote a chess engine[9] that can beat me if I'm playing fast but not if I try hard.
- Explored immediate-mode GUIs[10] for my chess engine and learned that oh, no, I really *do* love the web, thank you very much.
- Wrote a check-in post every day of batch that wasn't a holiday, and some that were. This was about 60 daily check-ins.
- Attended UTC-friendly check-ins every day at 8am Eastern. These were a *cornerstone* of my RC experience, because I got to have a call with the same folks (spread out all over the world!) every day, and it was such a good crew. I love you folks!
- Had about one coffee chat per day (sometimes they'd bunch up), so about 60 coffee chats across the batch! This was a wonderful way to get to know folks, share common interests, and hear about different life experiences.
- Read through most of the Red Book[11] and read a bunch of the papers from it, presenting them in the group.
- Wrote 25 blog posts, totalling over 32,000 words, during my batch. I got in the rhythm of writing and it feels profoundly good and correct.
- Pair programmed a lot, probably three times per week on average. I started out around once every day, but then tapered off near the end, so I think it rounded out to about this. I really loved the pairing experiences at RC, and I can't wait to keep doing it going forward!
- Worked through five chapters of Logical Foundations[12] with a few other brave souls. They've outlasted me. I threw in the towel, but had a *ton* of fun in the chapters we did work through. (We also tried learning Lean[13] first. There's a reason we switched to Coq.)

---

[7] https://softwarefoundations.cis.upenn.edu/current/lf-current/index.html
[8] https://github.com/ntietz/anode-kv
[9] https://github.com/ntietz/patzer
[10] https://github.com/emilk/egui
[11] http://www.redbook.io/
[12] https://softwarefoundations.cis.upenn.edu/current/lf-current/index.html
[13] https://leanprover.github.io/

- Wrote a (portion of) a chess database[14], which I'm still working on and is slow going! This project has been *very* fruitful and I've learned so much about systems programming, Rust, and databases.
- Worked through the first project in Crafting Interpreters[15]. The second one is still in progress.
- Learned a little bit of category theory, then let go of that pursuit because I was stretching myself too thin.
- Went to a few full-stack web development meetings, but let go of that mostly during batch since I wanted to focus on new-to-me things instead.
- Learned about homelab things with my fellow homelab enthusiasts!
- Bought and set up a very overkill homelab server, which then turned out to be *not at all* overkill for the chess database project.
- Did a few leetcode problems with people as a fun way to program together.
- Did the first few Advent of Code problems, and abandoned it when my batch ended as I was going to be short of time since I was going back to work.
- Attended the weekly Rust meeting! It was super fun to have a group of fellow Rustaceans to discuss things with, and it was a good way to measure my progress on getting comfortable with Rust. At the beginning, a lot of the discussion went over my head. By the end, I was pretty comfortable keeping up.
- Did the weekly reflections meeting every week and helped keep it running after the facilitator for the first half had his Never Graduation.
- Attended weekly presentations, and presented a few times on my projects!
- Switched my primary git forge to SourceHut[16] instead of GitHub.
- Learned some Idris[17] and decided it's not for me right now.
- Rediscovered the joy that is computers and writing code.

RC was pretty intense for me, in a delightful kind of way. Now it's time to find a sustainable balance.

30.3 What's next?

Well, what's next for me is that I've gone back to my day job, and I'm loving it. We've got a great team and a great culture. And I'm continuing on with Recurse Center. After your batch, you Never Graduate, and you

---

[14]https://sr.ht/~ntietz/isabella-db/
[15]https://craftinginterpreters.com/
[16]https://sr.ht/~ntietz/
[17]https://en.wikipedia.org/wiki/Idris_(programming_language)

stay a part of the community. I wrote most of this post while hanging out in a writing focus group with people from Recurse Center, and I'm so thankful for the community and being able to remain a part of it.

As for you: I'm not sure, but if you think Recurse Center sounds appealing, you should apply[18]. It's an amazing community of wonderful people, and it has been life-changing for me. It isn't *life-changing* for everyone (that would be quite a high bar!), but the experience is pretty excellent all around, and it's a great place to become a better programmer regardless of how much or how little experience you have.

If you join Recurse Center, welcome. I'll see you on Zulip.

And if not, no worries! Always feel free to reach out to me if you want to chat, whether that's Recurse Center or Rust or espresso or anything else!

---

[18]https://www.recurse.com/scout/click?t=c9a1a9e2e7a2ffefd4af20020b4af1e6

31 Reflecting on 2022, Looking Ahead to 2023

Originally published on 2022.12.28.

This is one of those cliched posts: Reflection on the year that's ending, and talking about goals and whatnot for next year. They're cliche, but they're also useful. The planning and reflecting process is a useful one, and sharing openly means other people can come along and learn with me.

31.1 Reflecting on the year

This year has been one hell of a year. I feel like I say that every year, but this one had way more in it than usual, or it feels that way.

Here are some of the highlights (or lowlights) in roughly chronological order. Of course, there's a lot more going on in my life than this, but I'm omitting family-related things.

I started the year in one of my deepest episodes of depression, and **recovered from it successfully** through a combination of therapy and a higher dose of my medication. This is the first time I've managed to use therapy as an effective tool, and it was tremendously helpful. I feel better equipped for the next time, and I suspect there *will* be a next time. I'm scared of it, because this one was *bad*, but I'm also more prepared for it than I was this time.

Russia invaded Ukraine, and I quit Twitter. This one I don't think needs a lot of expounding. It's been major news for the whole year, anyway. It hit me really had and I quit the last social media I was on (Twitter) as a result. I also had to stop reading a book shortly into it, because it was dwelling on a child dying.

My employer did a round of layoffs. I'm not going to share much about this publicly (not sure what I *can*, frankly, and I also want to keep my blog completely disconnected from work), except to say that the company has transformed into the ideal company for me at this stage. We have four-day work weeks now, and we also added a sabbatical program which I piloted. I still get some good technical challenges, but even more, we've leaned into the culture and flexibility that were keeping me there. For what I want to do right now, I cannot imagine a better place to be.

I did a 12-week batch at the Recurse Center. I've written about this previously in my return statement, but some bears repeating. It was a life-changing and formative experience. I went in expecting to be jazzed about databases and learn a lot. I did learn a lot about databases, but also a whole heck of a lot about myself. Which kinda leads into…

I use they/them pronouns now. I don't know fully what this all means, and I'm working on figuring out how this affects my life, or *if* it does. But I'm much more comfortable in myself now than I ever have been before. Also, my painted nails look *fantastic*. Going out for some glitter nail polish tomorrow to ring in the new year right. I haven't told a ton of people yet, so if you're reading this and a family member or friend: please reach out and say hi, I'd love to talk about it!

I wrote over 36,000 words on this blog. This was from 30 posts, including this one. This is more than I've written in any previous year. I used to write about 5,000 words per year across four to eight blog posts. The main thing is that I got into a good rhythm of writing at RC, and remembered how important it is to me. I got over my preciousness about my blog (not everything can or should be profound!), and in the process put out more good blog posts by releasing more in general.

The iconic mill in Kent, Ohio burned. This one is still a big open question for me, because a big part of it burned. What we don't know yet is the extent of the damage. Will the grain towers also have to come down due to the heat damage, or are they safe? This is an iconic building, and it was a big part of my experience in Kent. It was also one of the last remaining physical connections to my Grandpa Bill.

Inflation continued to rage. This is on the minds of probably everyone who has to work to make a living. (I did overhear a VC ask "Are people really feeling like they're earning less because of inflation?" Tell me you're a VC without telling me you're a VC.) This has an obvious cost to me, and it also makes everything feel different. The market is not running wild the way it was a year ago, so everything feels more constrained.

I'm sure I'm forgetting other thing that happened this year, but that's a lot of it as far as I can remember!

So… Lots happened, and it was a turbulent year. And yet, I come out feeling **much more stable in myself**. It's weird to say that when the world feels remarkably *unstable* right now, especially after having survived layoffs recently. But it's true: I'm more stable *in myself* and feel markedly more comfortable in myself.

During this year, especially during my time at RC, I really discovered who I am. I'm a software engineer, writer, and parent. I love all these in their own ways and they occupy different places in my life. I've let go of some other facets of identity; most notably, aspirations for building a startup.

31.2 Looking forward to next year

So, what will next year look like? Any predictions would be folly. But I've spent some time thinking about what I *want* next year to look like, and what I do or don't want to do. So, I have a few goals and more *non-goals* of things I explicitly want to *not* do.

I want to keep writing. I'm not going to set hard goals around word count or anything, but I do want to publish a blog post at least once every two weeks. That's 1/4 as many as I was putting out during RC, and I think it's sustainable. Hopefully I'll overshoot! My running has taught me that having a goal, though, having a plan, is essential for keeping forward momentum.

I will not put any side projects into production. One of my bad habits is starting on a project and letting it expand in scope until it has world-changing aspirations. This detracts from my learning and adds tremendous stress to the whole thing. Instead, I'll let each project serve its purpose for what I want to learn and write about, then let it go.

I don't want to learn about DevOps (on my own time). This was one of those things I learned this year. There's a lot there, and it's super important, but **it's not for me**. I'd like to spend as little time as possible learning about it (beyond what my job requires during work hours).

I want to stay active in the RC community. This community has been an instrumental part of my transformation this year, of my finding my footing and internal stability. I'm a better person for it. I'm privileged to be able to continue participating as an alum, and I'm going to stick around and keep working and learning and helping.

I want to establish habits for my learning. I've started on this already: I'm waking up earlier to slot in some time for programming or chess study before the kids get up. I'm going to make sure I find a rhythm that works for me (my first pass resulted in a sleep deficit) so that I can keep working on my programming, writing, and chess study.

I want to keep in touch with people. I have made so many new friends at RC, and I've also started reaching out to some old friends who I lost contact with. I've started setting up a habit process for staying in touch with people, and I want to stick with it. Social connections are important, and I can improve them with deliberate effort.

I think that's it! 2023 has the potential to be a fantastic year. Let's hope for a more peaceful, democratic, and healthy year than 2022. At any rate, I'll see you on the other side of New Years!

32 The back catalog

The following posts are my back catalog from before I started writing regularly. I would write when a Grand Idea struck me and I generally tried to have big ideas. It was the naivete of a fresh graduate who was always told she is smart, thrust into a world filled with many other smart people.

The essays here are often naive, but there are some gems. I think they're worth reading, but keeping them in their appropriate context as lightly-considered, somewhat naive gems of otherwise good ideas.

Consider, for example, the essay on burnout and running. When I wrote that, I was living with undiagnosed anxiety and depression. I was also extremely stressed and somewhat burnt out from a job, and had the eager earnestness of a young adult who has discovered a coping mechanism that works for her. The core message is one of hope: there are ways to cope, there are ways to improve your life, and it can get better. But it's definitely something I would write differently today.

I hope you will enjoy this back catalog, and that you will take away something of value. Please know that what is written here does not reflect what I think today, and so it's a historical curiosity.

33 The Beginning of Something

Originally published on 2015.02.22.

It seems like everyone in the software industry goes through a blogging phase. This is the beginning of mine.

I have started this blog time and time again over the last three years. My original inspiration for having a technical blog came from one of my mentors[1] at my internship. The continued inspiration is from people telling me that I sometimes make insightful comments.

This blog is not fully formed in my head yet, but I have some very broad topics that I want to address over time: data privacy, mental health, education, ethics, and life. I also intend to cover a smattering of technical topics. What I cover will certainly deviate from this, but it's somewhere to start.

---

[1]http://caseystella.com

34 How Cryptology Can Fix Identity Theft

Originally published on 2015.02.22.

Identity theft is a huge problem, costing Americans more than $4.5 billion in 2012[1]. Identity theft victims frequently lose time and money and undergo significant mental hardships while dealing with the fallout. It can happen a few different ways, but one large attack vector is through the identity verification process.

Every time your identity is verified, one of the following mechanisms is probably used:

- an array of challenge questions ("what were your last two addresses?")
- submitting a copy of a physical document (passport or id card)
- providing your Social Security number (SSN)

All of these come with problems. They are subject to two main attack vectors: social engineering, where a bad actor may trick you into giving up this information to them directly; or bad actors within a legitimate organization that you have to provide the information to. The second attack vector is far more insidious, since you cannot do anything to prevent it. If you submit your SSN with a form at your local community college and an employee handling the form copies it down, it is lost – but you had no choice and *had* to include the SSN.

Let's back up. What's the big problem here? Why are these mechanisms weak?

There are two classical problems in secure communications: authentication and encryption. *Authentication* is proving your identity. *Encryption* is protecting a message from all but the intended recipients. Together, these let you send messages which cannot be intercepted and can be demonstrated to be from you, not an impostor.

Traditional identity verification mechanisms are just means of *authenticating* your requests. These are based on shared information. Essentially, both Alice and Bob must have the same information to verify that Alice really is who she claims to be. Here's the problem: that means that Bob can then go to Mark and say "Hi, I'm Alice, here's proof!" and Mark would be fooled.

---

[1] https://www.fas.org/sgp/crs/misc/R40599.pdf

Solving this problem requires switching to an asymmetric information system. This is the same way that your bank's website proves that it is legitimate. A central authority, called the certificate authority (CA), issues a certificate to the bank. The bank holds private information it can use to sign a message (their private key), and then your browser checks the signature using the public certificate from the CA. No one else can impersonate the bank, because no one else has the bank's private key.

We can do the same thing for identity verification for people. With a central "Personal Identity Authority" (such a name evokes some dystopian imagery), we could issue every person a private and public key. The public keys would all be recorded so that anyone could see everyone else's public keys, but private keys would be held only be each individual. Then, identity proof would be done by a simple process. Imagine that Bob wants to verify Alice's identity:

1. Bob would send Alice a short message (randomly generated, and unique each time).
2. Alice would encrypt this message using her private key and send it back to Bob.
3. Bob would retrieve Alice's public key and use it to decrypt Alice's message.
4. If the received message matches the original one, then Alice is who she claims to be.

This system would be technically sound and would result in both far more secure identities and much higher confidence identity verification. However, it comes with problems of its own.

- Software systems would be necessary to implement the system. People can't encrypt random messages with large keys by hand. These systems are not awfully difficult to make (in fact, they already exist) but getting them integrated into everyone's phone, laptop, browser, and all the services they use, would be a significantly challenging endeavor.
- People would lose their private keys. If someone breaks their laptop or phone and their private key is lost, how would a new one be reissued? If you can use an old technique, like your SSN, to get a new key, then what would stop an attacker from simply pretending to be you and getting a new public/private key pair associated with your identity?
- People can have their private keys stolen. This could happen through security holes in their laptops and phones, or through social engineering to convince people to give up their private keys voluntarily.
- A great deal of trust is now placed in one central authority. This authority must be trusted not just to manage your identity, but also

to be responsible with a lot of information. All requests for your public key would be signals that you are authenticating in different places (Facebook wants your public key? That is a signal that you just used Facebook.), so the central authority would have a new wealth of tracking data.

I hope that within my lifetime, I can see symmetric information stop being used for identity verification. However, I also hope that these issues can be solved well *before* we implement any such system.

35 In Defense of the Midwest

Originally published on 2015.03.08.

As an undergraduate, I always imagined that I would someday move to the SF Bay Area to live in the heart of the software industry. With this in mind, in my final semester at Kent State, I joined a Silicon Valley startup as their third engineer[1]. The staff at that time was split: one founder and one engineer were in Mountain View, CA; one founder and one engineer were in Ohio; and one engineer was remote. Nearly every month in the first year, I flew out to the Silicon Valley office to work with the engineers out there.

Since then, we have grown to have a technical staff of about 20 people. We are split pretty evenly between the Silicon Valley office and the Ohio office. I spend most of my time in the Ohio office, but I do commute to the Silicon Valley one occasionally.

Nearly every time I go out to California, my coworkers ask me the usual question: "so, when are you moving to California?" It seems like for people in the Valley, moving to California is such an obvious choice that it isn't even a question of *if* I'll move, but *when*. However, I truly love the Midwest and that I want to stay here for as long as I can. It's not for everyone, but it is for me and maybe it is for you.

35.1 It is affordable

In San Francisco, the median one-bedroom apartment costs $3,120 / month[2]. In Kent, Ohio, a *really nice* one-bedroom apartment will cost you at most $1000 / month. Salaries are much higher in the Valley than in Ohio, but even a low salary in Ohio can get you a very nice apartment.

---

[1] My company is in stealth mode and has requested that we not talk about the company publicly at this time. When that changes, I might have more to say about the company. Note: I can talk to individuals one-on-one if anyone is curious.

[2] http://www.businessinsider.com/san-francisco-neighborhoods-where-one-bedrooms-are-expensive-2014-8

35.2 There is lots to do

Another argument that's used is that "Ohio is in the middle of nowhere" implying that there is nothing to do here and life is boring, surrounded by cornfields. On the contrary, there is actually a ton to do in Ohio. Here's a tiny sampling of what I like:

- Cleveland gets all the Broadway shows once they go off Broadway, but at about half the cost
- We have great music here, including the world-renowned Cleveland Jazz Orchestra, and tons of bands come through
- We have some great sports teams (hi, OSU) and a ton of great sports fans (hi, Browns fans)
- We have great food at very affordable prices

Actually, I haven't found anything I could do out in the Valley that I could not also do back in Ohio, except maybe get killer sushi.

35.3 Midwesterners are Great People

More than anything else, I love the people in the Midwest. Here's why:

- Our people are incredibly polite and caring. Out here, people greet you on the sidewalk even if they don't know you. Neighbors will come help push your car out of the snowbank it got stuck in. Cars will let let you merge when they don't have to.
- Our people are, well, scrappy: even though the Browns continue to lose, year after year, you will find no fans more loyal than the Browns fans[3]. This attitude is carried through most things we do: even if you fail over and over, you just keep trying and hoping.

On balance, I haven't found nicer people than in the Midwest.

35.4 The Weather

Not many people would claim that Ohio's weather is great, but count me among them. Our winters are fairly harsh and cold, but they make you truly appreciate spring when it comes. All the non-winter seasons are really nice: spring is pleasant and life is blossoming around you; summer is warm and laid-back; and fall is brisk and beautiful, with the leaves all changing colors.

---

[3]We have fans on every continent including Antarctica.

35.5 Great Universities

Despite popular opinion, the great universities aren't limited to the two coasts: we have UW-Madison, UIUC, Northwestern, and OSU, to name just a few. (Carnegie Mellon is also nearby, even though it isn't technically in the Midwest.)

35.6 The Pace of Life

On both coasts, the pace of life is really, really high: you just go, go, go and work constantly. If you go to a restaurant or coffee shop, people around you are probably all talking something work related, because people don't slow down very much.

In the Midwest, though, people take a much more relaxed pace. If you go to a coffee shop, you'll find people talking about real life things, not work. Maybe they're talking about a book they read in their free time!

This is one of the things I love most about the Midwest - people actually turn off work mode sometimes and go relax. I firmly believe that, even in spite of this, people are not less productive here than in the Valley, because even though we may put in fewer hours, those hours are more energetic and we are more recharged.

The Midwest is a beautiful place filled with beautiful people. Don't write it off just because it isn't the heart of Silicon Valley - there is still a lot of good stuff and good work being done in this part of the country. Come visit, stay for a while.

36 Book review: Data and Goliath

Originally published on 2015.07.13.

I just finished reading Bruce Schneier's latest book, "Data and Goliath." I was apprehensive at first – I'm a big fan of Schneier's posts online, but I found this randomly at the library and I was hoping not to be disappointed. In the end, it was well worth the read.

The book was split into three parts. In Part One, he discusses what a world of constant mass surveillance looks like. He illustrates what data everyone is leaking through ordinary activities, how people can and are monitored, and how this data can be used. In Part Two, he explains what is at stake: what the political and economic losses of surveillance are both in the US and abroad. And in Part Three, he explains what can be done about this in a three prong fashion: what the government should do; what corporations should do; and what we, the people, should do. All throughout, he provided compelling examples and illustrations, as well as footnotes with additional references (although, confusingly, these are not referenced inline but are merely listed at the end).

There were many compelling points in this book, and I can't list them here, but I want to call attention to one in particular. He puts out a call to action for the tech community to (paradoxically) create surveillance tools for the government to use - the argument being that "if we want organizations like the NSA to protect our privacy, we're going to have to give them new ways to perform their intelligence jobs".

Overall, I think he did a great job making these issues available for a non-technical audience. It was written in a way that will be open to everyone inside or outside the tech community. This book is a must-read in today's surveillance-filled world: buy it for your friends, get it from the library. Spread the word.

37 Fight burnout, go for a run

Originally published on 2016.02.19.

Here's something we don't talk about enough: burnout sucks and it can happen to any one of us. We need to talk about it. We need to know how to deal with it and recover from it. And we need to recognize that everyone can come back from it, stronger than ever.

In the software industry, we are subject to lots of pressure, long hours, and emotionally taxing work[1]. These are stressful and very difficult to deal with, and can ultimately lead to burnout. I hope those reading this have not had to deal with it, but I sadly suspect that most of you have (or will). I know that I have, and more than once.

The first time I experienced burnout was in 2012, when I ambitiously chose to take on not one, but two, undergraduate research projects simultaneously. Some people may thrive in this environment, but it ended up reducing me to tears and ultimately leading me out of academia and into industry[2].

The second time I experienced burnout was in 2014. I was working for a startup[3] which I believed in 100%[4]. Any job comes with stress and can have long hours (especially for a young software engineer who does not know how to set boundaries), and since I believed 100% in the company, I sacrificed too much while having no other outlet for my stress.

This stress culminated in me lying on the floor of my house, in tears, broken. Something had to change. Either I would find a way to deal with the stress, or I would have to find a new industry to work in, because this was simply not sustainable. Almost without thought, I walked over to the door, laced up my Nike running shoes, and went and ran the first damn mile I had run in a long, long time. The more my feet hit the pavement, the more my stress melted away, and by the time I was done I felt like a

---

[1] Personally, I find programming to be nearly constant emotional whiplash: the successes make me feel really great, and the roadblocks and failures make me feel really awful. Anecdotally, many of my coworkers have felt same way.

[2] I recently revisited this decision, and tried out grad school. Fairly quickly, I determined once again it was the wrong choice for me, but it was good to make this decision again when not experiencing burnout.

[3] http://www.graphsql.com

[4] It is still my belief that GraphSQL has essentially the best graph computing platform out there. Adam, please make a public release soon! I want to play with it again!

normal human being again. The repetitive, meditative nature of running melted it all away (and the endorphins didn't hurt, either).

It took me a long time to recover from burnout in 2012 and in 2014, but I did it, and I came back stronger for it each time. I now know how to set boundaries, how to relax and have a life outside of work, and I've adopted a hobby that will keep improving my health for a long time. And I joined a great company[5] doing some stuff I really care about, but I'm also really sure that the changes I've made will ensure I stay a strong, healthy engineer for years to come.

Unfortunately, my experiences here are not unique.

If you are fighting with burnout right now, please, join me: go for a run, or a bike ride, or a long walk (no phone allowed). Your mind will be clearer[6] and it will be one small step on the road to recovery (if not, at least you still got some exercise!). (And if you need someone to talk to, get in touch.)

If you have gone through burnout or fought with mental illness, I beg of you: post your story and share it wide. As an industry, we have a responsibility to protect each others' health, physical and mental. A big part of that is sharing our stories and our coping mechanisms so no one has to feel alone, trapped, or hopeless.

---

[5] http://crosschx.com/
[6] http://news.stanford.edu/news/2014/april/walking-vs-sitting-042414.html

38 Surveillance, schools, and our children

Originally published on 2016.03.07.

In 2010, the news broke that Harriton High School, in a suburb of Philadelphia, was activating webcams on student laptops[1][2]. When they were at home. **In their bedrooms**. They captured photos while students were in private spaces, where they never expected to be watched.

A few days ago, I heard about another school that is also surveilling their students: AltSchool[3]. They are taking a very different approach: the cameras are visible and are there to help improve education, to conduct research and find out how to more effectively educate our students.

On the face of it, it looks like AltSchool is doing something noble. At the very least, the surface level does not appear to be immoral, let alone nearly as repugnant as what was done at Harriton High School. And in some ways, that is true: the surveillance itself does not appear to be leading to negatives here, and it passes the minimum bar of informing all involved parties.

However, there is an insidious side effect, and one which is far worse: it will acclimate the students to a surveillance state. In our society we are fighting a battle for our privacy right now, and in many ways, the next generation will be the one to seal the deal. Either they will embrace and extend privacy tools and policy, or they will embrace and extend government and corporate surveillance. By exposing our children to pervasive surveillance during their most formative years, we risk permanently shifting the balance toward surveillance and numbing our children to its dangers.

Don't get me wrong, I think that there is room for massive improvement in education. However, the solution ought to include positive new technology, like better adaptive learning[4] tools. We are better than this. We can innovate and create great new tools that will help, or even revolutionize, our children's education. But if we want to do that, we need to

---

[1]https://web.archive.org/web/20160113065213/http://www.huffingtonpost.com/2010/02/22/harriton-high-school-admi_n_471321.html

[2]More information is available on Wikipedia (https://en.wikipedia.org/wiki/Robbins_v._Lower_Merion_School_District#Covert_surveillance).

[3]http://www.newyorker.com/magazine/2016/03/07/altschools-disrupted-education
[4]https://www.duolingo.com/

do it ethically and ensure that we do not accidentally harm society while trying to help it.

The ACM has a code of ethics[5] for software engineers. From it: *"Approve software only if they have a well-founded belief that it is safe, meets specifications, passes appropriate tests, and does not diminish quality of life, diminish privacy or harm the environment. The ultimate effect of the work should be to the public good."*.

I welcome you to think and comment about this: if we subject our children to pervasive surveillance, will that lead to less privacy? Is that to the public good?

---

[5]http://www.acm.org/about/se-code

39 Book review: The Circle

Originally published on 2016.03.15.

Surveillance has gotten a lot of media attention lately (and a bit of attention on this very blog), and for good reason. So, it should be no surprise that it's also turning up in our dystopian novels!

"The Circle" is a dystopian novel by Dave Eggers. While fiction, it is set in a plausible universe which is alarmingly similar to present day, and it lays out a future which we could slide into if we are not careful about corporate and government surveillance. Eggers' message of the dangers of surveillance is both clear and harrowing.

I would strongly recommend this book to any of my friends, and I would make it required reading for technologists. As technologists, we hold the keys to either a great or terrible future, so we must together carefully weight the future we are creating.

40 Starting a New Chapter

Originally published on 2016.08.21.

At the end of this week, I am starting a new chapter of my life: entrepreneurship. This is my last week at CrossChx[1], and then I begin splitting my time between contract work and developing some of my own ideas.

I only spent about three quarters of a year at CrossChx, but in that time a lot has happened. I've made some friends who will be with me for a long time. I've written some code that I'm really damn proud of. And I've learned a lot about what I want in life. Right now, what I want in life is the freedom to pursue my dreams, the freedom to make the things that I really care about, the freedom to leave a Nicole-shaped dent on the world.

One of my colleagues requested that I start a blog (I already have one) around my adventures through this new chapter. Other coworkers have asked me to give them tips and updates since they have considered making similar moves. Here it is. I'm going to try to maintain weekly updates (ideally on Fridays) talking about my experiences with both contracting and entrepreneurship / independent development. It isn't really clear what this chapter is going to look like, but it sure will be interesting!

Wish me luck!

---

[1] http://crosschx.com

41 Consider part time work

Originally published on 2016.09.26.

It has long been predicted that with more automation and more technology, we could all work less and have more leisure time, but we continue to fall short of that promise. In many ways, we're working harder and longer, with more stress, than previous generations did. I think that a large part of that is because of societal pressures to work long hours, even when doing so doesn't make sense.

That doesn't make a lot of sense to me. We shouldn't work long hours just for the sake of it, especially because right now, conditions are almost perfect for accommodating part-time work. It would benefit everyone if we could reverse societal pressures and encourage part-time work and shorter hours.

Many arguments center around the benefits to the employees - which are numerous - but there are immense benefits to employers, families, and society as a whole.

41.1 Employers Benefit

Here are a few of the reasons that you should consider part-time work for your employees, whether you're running a startup or a multi-national corporation:

- Programmers are like cows... and we know that happy milk comes from happy cows. Traditionally, we have tried to make programmers happy by giving them perks like ping-pong tables and free beer, but those are exclusionary perks (not everyone drinks, not everyone wants to live like they're in college) that benefit a particular demographic, whereas everyone can benefit from having fewer hours, so they have more to do what they want.
- You remove waste while retaining throughput. When you cut down hours, you will mainly remove the wasted hours - those spent on long coffee breaks and long lunch breaks and talking at the water cooler. But you will also remove other forms of waste, as your employees will start to police meeting length and cut down on the Nerf dart battles, because when their time is limited they will not want to waste it.

- You can retain people who would otherwise leave. There are plenty of people who leave jobs because their schedules aren't flexible enough. I recently left a job because the hours were not conducive to the other projects I want to work on, and it's fairly common for new parents to leave for a job that gives them more time with their families. If you let employees work part-time, you will be able to retain these talented employees and avoid leaving gaps in your lineup.

41.2 Employees Benefit

The benefits to employees are fairly self-explanatory, but here goes anyway:

- You get to have a fresher mind when you're at work, because you're not at work as often.
- You have more free time to explore the hobbies you love, whether that is running or reading or even more programming.
- You will cut down on wasted time at work (less reddit, shorter meetings, shorter coffee breaks) and will end up leaving feeling much more fulfilled.
- For those of you with families or planning on having one, you get more time with your family - who can argue with that?
- If you want to start your own company, it gives you another option instead of just quitting your job or trying to burn the candles at both ends while working a full-time job. (This is what I'm doing - consulting part-time, and starting a company in the rest of my time, so I work full-time but only get paid for part of it.)

41.3 Society Benefits

Probably most important here are the benefits to society at large, especially because getting a lot of part-time work will require a societal shift so that it is not looked down upon to avoid full-time employment (and to pressure employers to allow it and provide benefits to part-time workers). Here are just a few, although there are many more I've missed:

- We can create more white-collar jobs. There is a certain amount of demand for programmers and accountants and actuaries, so if each of these employees provides fewer hours, we can hire more of them. In the long-run, this demand will encourage creating job training programs, encourage more people to pursue these fields, and hopefully help elevate more people to the middle or upper class.

- It makes for a more equal society and reduces some gender barriers for women (and men) who want to be parents. No one should have to choose between having a career and being the primary parent, so accepting part-time work would aid in this. Primary parents could still have careers. Children of career-oriented people could still have parents. Everyone involved gets more time with those they love and it would be great.
- Society would have more innovation and more startups. As other nations are getting more and more innovation, the US is at a critical juncture. We need to ensure that our economy stays strong and our innovation sector stays at the forefront if we want to remain economically competitive - let alone dominant - for years to come. People come up with their best, most innovative ideas when they are well rested and when they have time to just sit and think and be bored, so let's create more of that. Having longer hours and longer commutes may grind out productivity right now (although, I'm skeptical) but in the long run it will not benefit our society. We need a culture that fosters creativity, not grinding out widgets.

So, that's my pitch. I think that any of you who want to be more creative, who want to learn more, who want more freedom - you should consider working part time, and you should consider the same for your employees. I've taken the leap, and so far it has been great. I'm more creative than I was a month ago, and it seems like I'm becoming more creative every day. Join me.

42 Security of the Infinity Ergodox on MacOS

Originally published on 2016.10.12.

A friend of mine is very into keyboards and, after seeing his keyboards at work and admiring his Ergodox many times, I took the plunge and built my own. 152 solder joints later, I have this beauty:

It took a few days to get used to it and in the process, I found a bug in layer switching[1], which I contributed a fix for[2]. While fixing it, I came

---

[1]https://github.com/kiibohd/controller/issues/66
[2]https://github.com/kiibohd/controller/pull/156

across some very cool debugging features[3] - the keyboard has a console which gives debug info and is very easy to connect to:

```
screen /dev/tty.usbmodem1A12144
```

This console gives a lot of debugging information, and it turns out that it can show every key press! Neat, until you realize that *any* user of the system can also see every single key press. A non-privileged test user on my Mac[4] was able to read every key press I made while typing as my normal user.

This is a huge breach of security. I routinely create accounts on my desktop for other people (my fiancée, my friends who are learning to code), so this is simply an unacceptable risk. This is present in keyboards built with custom firmware, but also on the firmware that ships with the keyboard or is downloaded from the online configuration tool[5].

Fortunately, the firmware was created to be pretty modular, and it is easy to turn this functionality on or off by adding just a few define guards:

```
#if defined(DEBUG)
// Enable CLI
CLI_init();
#endif

// ...

#if defined(DEBUG)
// Process CLI
CLI_process();
#endif
```

What this does is turn off initialization and processing of the CLI. It is still there, sitting in the background - and there might still be more security risks with it - but the obvious attack vector is gone.

On October 10, 2016, I submitted an issue[6] to address this, and a corresponding pull request[7]. Following a discussion with Haata (the maintainer of the firmware), we decided to pursue adding an option to have

[3]https://github.com/kiibohd/controller/wiki/Debugging
[4]I verified the issue exists on OS X, but it does not exist in Linux since you need root access to access the console. However, the documentation suggests adding a udev rule file which does give read permissions to everyone without sudo, so many Linux users are likely vulnerable.
[5]https://configurator.input.club/
[6]https://github.com/kiibohd/controller/issues/159
[7]https://github.com/kiibohd/controller/pull/160

a security-hardened mode, as well as adding a passcode to enable to console on non-hardened keyboards.

My personal recommendation is to apply my patch to your firmware if you are using OS X[8]. On Linux, you shouldn't have to patch anything immediately, since accessing the console requires sudo permissions.

Stay posted for more updates! I hope to have the first pass at the security-hardened mode out during October, and hopefully the corresponding configurator changes can follow shortly after.

---

[8]You can get the patch from my pull request, it works and is only closed because it is not the long-term solution. I'm using it myself for now. If you need help, email me or tweet at me.

43 Functional programming and big data

Originally published on 2016.11.12.

This post is a long one, so here's a brief roadmap. We'll start with a quick introduction to functional programming. Then you'll get a quick introduction to Apache Spark and the history of big data. After that, we will get to a hands on demo of Spark. Okay, are you with me? Let's go!

43.1 Intro to Functional Programming

43.1.1 Motivation

First of all, why should you even care about functional programming?

Simply put, functional programming matters because it is a big part of the future of the software industry. The industry is buzzing about functional programming (FP). Elements of FP are working their way into most mainstream languages. Even C++ and Java, stalwarts of the procedural object-oriented camp, have adopted lambda functions. It is less common to see FP adopted wholesale, but functional languages like Scala, F#, and Clojure are gaining in popularity. Although uncommon, companies are even using Haskell in production systems[1].

You should care about functional programming even if you never use it in production (although, I suspect you will). Functional programming gives you a completely different way of thinking about problems and is a good tool in any programmer's toolbelt. Of course, getting this other perspective comes with a price: FP usually takes a significant investment to learn and to learn well.

43.1.2 Fluffy Abstract Explanation

So, with the benefits in mind, let's tackle the first question: what *is* functional programming? Wikipedia defines it as "a programming paradigm [...] that treats computation as evaluation of mathematical functions and

---

[1]https://www.wired.com/2015/09/facebooks-new-anti-spam-system-hints-future-coding/

avoids changing-state and mutable data". Let's break that down piece by piece:

- **a programming paradigm** is a essentially a style of programming and the features it uses. Paradigms you'll hear about most frequently are: imperative; object-oriented; procedural; functional; declarative. There is often overlap between these, and it's mostly a way to classify languages and talk about them more easily.
- **computation as evaluation of mathematical functions** means that instead of a "data recipe" where you have a set of instructions that you follow step by step, you describe with math what you expect as output based on what you provide as input. That is, you precisely describe the relationship between the set of all inputs and the set of permitted outputs of your function.
- **avoiding changing-state and mutable data** means that you can't say x = 5 and then later say x = 10. When you set a value equal to something, it is equal forever and you can't change the state. If you create a list and you need to add a new element to it, you don't modify it in-place - you create a new list with the element added to it. This gives a few nice properties: you don't have to worry about concurrent accesses to data structures, since those are read-only; you don't have to worry about a function modifying data you pass in, since it can't; and it simplifies testing.

So, in a functional programming language, you write code using functions that don't have side effects. Since we are arguably removing features from imperative languages (mutable data, side effects, etc.), we must also be adding features (or creating a very strange language). Here are a couple of features we will always have in functional languages:

- **Higher order functions**: functions that can take functions as arguments, and can return functions as results. This makes it so you can do really cool things like writing your own control structures. We'll see examples of this in the next section, since it underpins most of functional programming.
- **Lambda functions** are anonymous functions. They sometimes have restrictions in what they can do (for example, lambdas in Python cannot do everything lambdas in Haskell can do) but in principle, a lambda function is just an unnamed function.
- **Algebraic datatypes** are composite types, most commonly product types (such as tuples or records) and sum types (such as union types). We will also see examples of these in the next section.

There are a lot of other features that you see more in functional programming languages, but it is important to keep in mind that not all FP languages are Haskell, and you can do FP even if your language is technically in a different paradigm (for example, JS has a strong community building around doing FP, especially with the rise of frameworks like React[2] and Redux[3] and libraries like Ramda[4]).

43.1.3 That made no sense, show me the code

Let's not pretend that that was perfectly clear. Unless you've actually done some functional programming, that explanation is likely abstract and not perfectly clear, so let's look at a few concrete examples. These will all be in Scala (it can show both imperative and functional styles, and it is the language used for Spark).

43.1.3.1 "Hello Fibonacci"

The canonical example for getting started with functional programming seems to be calculating the Fibonacci sequence. It's short and digestible and shows a little bit of the flavor (and avoids IO, which can be difficult in functional languages).

n.b.: I'm assuming the user will *always* pass in valid input, and we aren't concerned with error handling here. That's for another blog post.

First, let's take a look at an imperative implementation:

```scala
def imperativeFibonacci(n: Int): Int = {
  var a: Int = 0
  var b: Int = 1
  var index: Int = 0

  while (index < n) {
    index += 1

    val next = a + b
    a = b
    b = next
  }

  a
}
```

This is basically the version we all wrote when we were learning. It was kind of tricky to write, and a lot of that trickiness comes from the fact

---

[2] https://facebook.github.io/react/
[3] https://github.com/reactjs/redux
[4] http://ramdajs.com/

that when we look at the definition of the Fibonacci series on Wikipedia[5], it is not expressed as this kind of calculation. Wouldn't it be nice if we could write it in a way that's closer to how it's defined?

We're in luck. Here is one way we could write a functional implementation:

```
def fibonacci(n: Int): Int = n match {
  case 0 => 0
  case 1 => 1
  case _ => fibonacci(n-1) + fibonacci(n-2)
}
```

This is much cleaner. It has two major problems, though: it will result in a stack overflow if we run with too high of an n value, and it will be really slow for large n (it's O(2^n), which makes kittens cry).

Here's another functional approach which is still clean and avoids both of these problems:

```
def fibonacci(n: Int): Int = {
  def fib(n: Int, a: Int, b: Int): Int = n match {
    case 0 => a
    case _ => fib(n-1, b, a+b)
  }
  fibHelper(n, 0, 1)
}
```

This one avoids stack overflows by using tail calls[6], which are optimized by the Scala compiler and turned into loops. It also is more efficient, since it compiles down to something very similar to our imperative version above.

What makes this better than the imperative approach? Truthfully, it isn't necessarily better. It *definitely* is different, and having a different approach will benefit you.

43.1.3.2 Examples (Lambdas, Maps, Folds)

Now we have seen a basic example, we should look at a more thorough, complete, and realistic example. This is obviously contrived, but it should give you the flavors of functional programming.

---

[5]https://en.wikipedia.org/wiki/Fibonacci_number
[6]https://en.wikipedia.org/wiki/Tail_call

Let's pretend that you're a professor and your program has a list of student records in it (containing name, id, and grade). First, let's define the datatype we are using:

```
case class Student(name: String, id: String, grade: Float)
```

Now you want to know who is failing your course so you can intervene and help them get a better grade. We need to find the students who are currently failing. As an imperative programmer, you might write something like this:

```
def getFailingStudents(roster: Seq[Student]): Seq[Student] = {
  var disappointments = Seq[Student]()
  for (student <- roster) {
    if (student.grade < 90.0) { // we have high standards
      disappointments :+= student
    }
  }
  disappointments
}
```

If you also want to find the students who are passing, you will have to write nearly identical code. Let's see how we would do both of them in a functional style. I'm going to skip actually implementing the filter function and just show you how we do it with some functional constructs (higher order functions, lambda functions):

```
val failingStudents = roster.filter(x => x.grade < 90.0)
val passingStudents = roster.filter(x => x.grade >= 90.0)
```

Without higher order functions, we would not be able to define this kind of filter function. (We could hack it together using anonymous classes and overriding methods, like was done in Java for a long time, but that is ugly and very cumbersome; this is very clean.) The great thing about doing filters this way is we don't have to reimplement anything for passing students, we just use a different predicate.

Now let's compute the average grade of your students. Again, first imperative...

```
def averageGrade(roster: Seq[Student]): Seq[Student] = {
  var total = 0.0
  for (student <- roster) {
    total += student.grade
  }
```

```
  total / roster.length
}
```

...and then functional...

```
val sum = roster.map(student => student.grade)
               .foldLeft(0.0)((a,b) => a + b)
val avg = sum / roster.length
```

Here we have introduced two new concepts:

- `map` is used to transform one list into another list. It applies the supplied function to every element of the list. In this case, we transform a `Seq[Student]` into a `Seq[Float]`. This generally preserves the *structure* of the list, but transforms the *content* of it.
- `fold` is used to compact down a list and generate a resulting value (`foldLeft` and `foldRight` just control associativity[7]). The first argument is the initial accumulator, and then it applies the given function to the current accumulator and the next element of the list to generate the new accumulator. In our case, we transform a `Seq[Float]` into a `Float` by summing up the list. Note: `fold` is also sometimes called `reduce`.

43.1.3.3 What's Left?

There is a wealth of knowledge out there to gain in functional programming, and this introduction has come nowhere close to telling you everything useful about it. All of you should spend some time on reading and learning about functional programming. Hopefully, this has been a useful taste and will give you at least some value. Now we have to move on to other things.

43.2 Why is FP in Big Data?

I think at least a little bit of the hype about functional programming lately is thanks to the big data community. That should be apparent after learning more about how it is applied. Let's go through the history of big data to see how we've gotten to where we are, then go through the core concepts from FP that are useful in big data and how to use them and apply them.

---

[7] https://en.wikipedia.org/wiki/Operator_associativity

We haven't always had the infrastructure needed for handling big data, in terms of network speed and storage capacity. One of the first companies which had both the capacity for big data and the need for it was Google[8]. Another was Yahoo. (It turns out, the internet is *big* and generates a lot of data.) One of Yahoo's search engineers, Doug Cutting[9], created Lucene in 1999. The project ran well for a while but was running into a few problems, and Google happened to release a relevant paper on a distributed filesystems, which was then integrated into Lucene. Again in 2004, Google released a paper about a framework called MapReduce, and then it was integrated into some of Yahoo's infrastructure. In 2006, this integration was pulled out into its own project, called Hadoop[10]. The Hadoop ecosystem grew over time and eventually some very smart folks at Berkeley created Spark, which is basically the de facto big data processing framework now.

So, what is MapReduce, and what is Spark?

What is MapReduce? Simply put, *MapReduce* is a way to compute on large amounts of data by providing `Map` and `Reduce` operations. You can have as many iterations of your computation as you want, and in each one, you define a `Mapper` which is run over each input record and generates output, and you define a `Reducer` which reduces down the results and either prepares them for output or for further computation. These operations are designed to be run across many machines, often hundreds or thousands, so we have some specific requirements we need to support that. We discussed `Map` and `Reduce` (`fold`) above, so we already know that these concepts are drawn from functional programming. It's curious that the entire computing model Google released is based around two fundamental functions in functional programming, so we have to dig in to see *why* those functions were chosen. It turns out that the assumptions we make for functional programming are very helpful in doing distributed computations:

- **Avoiding side effects makes life better.** With functional programming, one of the core tenets is that you do not use side effects when computing values, so if `f(10)` returns 3 the first time you evaluate it, then `f(10)` will return 3 every time you evaluate it. Why does this matter for distributed computing? Because machine and network failures are fairly common, and you are almost guaranteed to encounter them when you run a cluster of hundreds or thousands of machines. If your computation always returns the same output for the given input, then dealing with failures is easy - just rerun the failed part of the computation on a new machine. But if it doesn't always return the same result (such as doing a distributed random

---

[8] http://lmgtfy.com/?q=Google
[9] https://en.wikipedia.org/wiki/Doug_Cutting
[10] https://en.wikipedia.org/wiki/Apache_Hadoop

shuffle of an array), then you have to start the entire computation over if any single part of it fails.

- **Avoiding global state makes life better.** This goes hand-in-hand with avoiding side effects, but is a subtly different point (or a more specific one). By avoiding global mutable state, you make it really easy to distribute your computation across many machines, because you no longer have to worry about shared global locks or synchronizing state between the machines. You only have to worry about getting each machine the data it is computing on.
- **Without side effects, testing is easier.** Since our computation doesn't (or shouldn't) have side effects, we can test things more easily, because we don't have to reset the computation between runs. We just pass in reasonable input to the test and as long as we get back the correct output, we are good to go. Whereas with side effects, we would have to worry about cleaning up after the tests, make sure that the computation can run correctly even if a previous run failed, etc.

Now, Hadoop (the open source implementation of MapReduce) was not perfect. Since Java did not support lambda functions or first-class functions until very recently, Hadoop MapReduce required you to write classes for the mapper and reducer, and these were very large and very clunky even when you were doing something relatively simple. Some people figured the solution was to add bindings for Python, where these implementations could be much shorter. However, it is still a big lift to write a *class* in order to just run a couple of *functions*... we should be able to pass those in directly. Further, people started to recognize that MapReduce was not the perfect paradigm for solving every single problem - it worked very well for some, and most could be shoved into it, but it wasn't perfect.

Along comes Spark to save the day.

What is Spark? Apache Spark[11] is an engine for large-scale data processing. It lets you do things like compute product recommendations, figure out duplicate patients in an elecronic health record system, and analyze clickstream data for that sweet, sweet advertizing revenue. Basically, it lets you pump in a lot of data, do some computations on it, and pump out results (and supports doing this on streaming data, too). This is a lot like Hadoop MapReduce, except that you are not restricted to running a map and a reduce over your data - you can do many other operations. All of this was enabled by the work done on Hadoop, which was generalized into a resource manager[12] which Spark was later written on top of.

---

[11] https://spark.apache.org/
[12] https://hadoop.apache.org/docs/r2.7.2/hadoop-yarn/hadoop-yarn-site/YARN.html

So, if we can do the same things we could with Hadoop MapReduce, why do we need Spark at all? Well, we need it because it borrowed more from functional programming - and being written in Scala, these functional concepts are much easier to apply.

- **First-class functions make life easier.** Instead of defining a mapper class, we just pass in a mapping function: `ourData.map(_ + 1)`. Instead of taking another whole file for the class just to create a function to pass in as the mapper, we can do it in one line, by just defining the map function.
- **We get better error handling.** Instead of returning `null` when a computation returns nothing, or manually crafting a datatype we can return that captures either-this-or-nothing, we have built-in datatypes that cover this (`Either` and `Maybe`), and we get an added bonus - any code that pattern matches against our return type is forced by the compiler to handle both cases, so we can rest assured that we won't have unhandled code paths. This is mostly a benefit brought in by algebraic data types.
- **Operating over collections is easy.** Remember that filter example above? We can do exactly that in Spark by just passing in a filter. The same with averages, or any other computation we can think of. Spark exposes a collections API we can use much like the built in collections, so we can do things almost exactly like we would on in-memory data (in a functional style), and get distributed computation for free.

43.3 Hands on with Spark

Now that we've learned what Spark is and where it came from, let's get our hands dirty with some actual examples of how Spark works. We will look at some standard functional programming functions and properties, and how these apply to writing Spark jobs.

43.3.1 Higher Order Functions

Now let's go through some of the common higher order functions you'll use when you're writing Spark jobs.

43.3.1.1 Filter

In functional languages, filtering lists (or any collection) is simple:

```
List(1,2,3,4,5).filter(x => x%2 == 0) // Only even numbers
```

We can do the same thing in Spark:

```
rdd.filter(x => x%2 == 0) // Only even numbers
```

It is the same operation we had before. We simply pass in function, and it gets applied to our data automatically.

43.3.1.2 Map

Mapping over a collection is a way of converting a collection of one type into a collection of another type. Suppose you have a list of `String`s:

```
List("1","2").map(x => x.toInt)
```

We can do the same thing in Spark:

```
stringNums.map(x => x.toInt)
```

The problem here is that sometimes we might have something that cannot be parsed, and Spark will abort the job if it fails too many times, so we should not have uncaught exceptions. How do we solve this problem in a functional style? We simply use the `Option` type (and a handy Scala wrapper that turns exceptions into `None` and returned values into `Some(...)` values). Here's the same conversion, but safe:

```
stringNums.map(x => Try(x.toInt).toOption)
```

This is great, but the problem is we now have `RDD[Option[Int]]` where we wanted `RDD[Int]`. How do we correct this? By reading the next section!

43.3.1.3 Flattening

When we have a list of lists (or generally, a collection of collections), we can *flatten* that into just the outer shell. Essentially, we take the innermost nested elements, and we pull them out of their containers into the parent containers. That's kind of hard to understand abstractly, so let's look at an example. Here's some vanilla Scala code that takes a `Seq[Seq[Int]]` and applies `flatten`, resulting in a `Seq[Int]`:

```
Seq(Seq(1,2), Seq(3), Seq(), Seq(4,5,6)).flatten == Seq(1,2,3,4,5,6)
```

We can do this with `Options`, too! Here's what that looks like:

```
Seq(None, Some(1), Some(2), None, None, Some(3)) == Seq(1,2,3)
```

Okay, so now we need to see how to do it in Spark. Spark, unfortunately, does not have `flatten` built in, but it does have `flatMap`, which means "apply map to this, and then flatten the results". We can work with that. There are two ways we can rewrite our old code to utilize our newfound flattening capabilities:

```
stringNums.map(x => Try(x.toInt).toOption).flatMap(identity)
stringNums.flatMap(x => Try(x.toInt).toOption)
```

The first line maps over the collection and then flattens it after the fact, while the second just uses `flatMap` in the first place and flattens it as it goes. The second is preferred, but the first is an option if you have a really good reason to do it.

43.3.1.4 Reduce (and friends)

We saw reduce before, and we can use it in Spark, as well. Let's say we have an `RDD[Student]` that contains all our students, and we want to compute the average grade right now. We can do that by first extracting their grades, then reducing across it, and then dividing that by the total number of students.

```
val numStudents = students.count
val sum = students.map(s => s.grade)
                   .reduce(_ + _)
val average = sum / numStudents
```

What if we want to count the words in a document? Suppose we have the document line-by-line. Then we can use one of the cousins of `reduce`, `reduceByKey`, to do this after we turn each word into a word-count-pair. This example leverages `flatMap` and `map`, and then combines everything down with a `reduceByKey`:

```
lines.flatMap(line => line.split(" "))
     .map(word => (word, 1))
     .reduceByKey(_ + _)
```

At the end, we will have turned an RDD[String] into RDD[(String,Count)] and we have the word counts we were looking for.

There are other higher order functions we can also use, and these are available in the API docs[13]. Now, let's move on and look at a couple of other things we need to know about how functional programming applies to Spark.

43.3.2 Associativity and commutativity

First, some super dry terminology:

- An **associative** operation is one where you can add in parentheses wherever you want and still get the same result. This means that, to be associative, we must have: (a + b) + c == a + (b + c). This holds true for most things we do, like addition and multiplication, but does not hold true for exponentiation: it's not the case that (2 ^ 3) ^ 4 == 2 ^ (3 ^ 4). It's also not true that (2 - 3) - 4 == 2 - (3 - 4).
- A **commutative** operation is either one that drives to work, or it's one where you can rearrange the order of the elements and still get the same result. This means that, to be commutative, we must have a * b == b * a (note: the * can mean multiplication, but it stands in for any operation we are doing). So, we can notice again that this does not hold for exponentiation or subtraction, but does hold for addition and multiplication.

This is important to understand when writing Spark programs, because you need your operations (usually) to be associative and commutative. If they are not, your code will have race conditions and non-deterministic behavior, and may also crash Spark.

Suppose you wrote this:

```
someNumbers.reduce(_ - _)
```

What would you expect the result to be? The short answer is: we don't know. Since the operation is not associative and is not commutative, we have broken *both* constraints we need to have this operation work well. In practice, this will probably kill your Spark job and will definitely give you unpredictable results if it *does* finish.

Usually you won't try to reduce with - or ^, but this is something to keep in mind always. I know from personal experience that with sufficiently advanced Spark jobs, you can break associativity and commutativity in

---

[13]https://spark.apache.org/docs/latest/api/scala/index.html#org.apache.spark.rdd.RDD

subtle ways that will eventually come out but be very difficult to debug. So keep it in mind, and think about this if your job sporadically fails.

43.3.3 What if you try side effects / IO?

Another thing to note is that sometimes, it is tempting to do IO or side effects within your Spark job. For example, you might want to compute a new interest rate for each customer, then write it back to the database:

```
customers.map(cust => calculateNewInterestRate(cust))
        .map(writeToDb(cust))
```

The problem is, we've just massively distributed our computation, and now we are going to essentially do a distributed denial of service attack on our database! This is problematic for obvious reasons, and I'd say that folks wouldn't try this, but I've seen it done, at places I've worked or where friends have worked.

You can also do something similar by reading data in, such as configuration files:

```
customers.map(cust => if (getConfig.flagIsOn) .......)
```

If you aren't careful, you'll read the configuration file for every single customer, and then your operations team will come hunting for you. Let's hope they don't have any unresolved anger issues.

Beyond just having your ops team hate you, this style of coding also is very difficult to test, because you have to have the configuration server/files, your database, etc. available just to run the code, even if you're not testing that interaction.

So, how do you resolve both of these cases? Basically, you do what you are supposed to do in any functional programming language: cleanly separate anything that is "pure" (no side effects, no IO) from anything that relies on the outside world.

43.4 Conclusion

Hopefully by now, you have the basic flavor of functional programming and you've seen how it has influenced Spark, and big data in general. There is a lot here to learn, but it is worth it and will ultimately make you a stronger engineer by giving you a second, independent way of thinking about your problems.

43 *Functional programming and big data*

If you have any questions, feel free to contact me (info in the side bar).

44 Growing teams and baking bread

Originally published on 2017.01.21.

One of the keys in baking bread is getting the dough to rise well. As the yeast does its work, it ferments some of the sugars in the dough into alcohol and carbon dioxide, resulting in a growing, bubbly mass of dough.

There are some tricks to making dough rise quickly, like using more yeast, using instant yeast, or even with a microwave[1]. These are methods of convenience, because they let you get the finished product out the door more quickly so you can eat your delicious bread.

But how do you make truly great bread? One of the ways to make a great bread is to give it a much longer time to rise. With a quick rise, a lot of the flavors are underdeveloped. For a simple sandwich bread, that might be okay. But for an artisanal crusty loaf, these flavors lend complexity of flavor and depth of development which is key. Those flavors come from having a long, slow rise, where the yeast can take its time fermenting and the flavors can develop, lending subtleties and complexities. A long rise also helps with good gluten formation, where the yeast will develop it naturally instead of requiring a lot of kneading to force everything into line.

The same is true with growing a team.

You *can* grow teams quickly, but by doing so, cohesion doesn't happen naturally and you have to force it, and the team culture that forms isn't as natural as the team culture if you grow a team slowly over time.

In contrast, if you grow a team slowly and organically over a longer period, you reap a lot of benefits. The team works out a lot of problems with cohesion naturally over time (instead of in rapid, very painful periods) and they will all grow together, leading to a very strong shared culture with similar values and similar goals.

There are definitely some situations where rapid growth is needed or beneficial, but it is worth thinking about whether or not it is necessary. A long, slow rise can make a unique team that has strong cohesion, and a more sustainable one at that.

---

[1] http://www.thekitchn.com/proof-your-bread-dough-in-the-microwave-35685

45 PyOhio

Originally published on 2017.07.30.

This was my first time going to PyOhio, and it was a blast. There will be some videos being posted soon, so I will opt to link to those as they come in, but first, here are some of the highlights:

- Ed Finkler of OSMI[1] gave a great talk on mental illness in tech, resources that are available, what OSMI does, etc. This topic *needs* to be discussed more (and I will have a very personal post about it myself coming soon). If you have the means, please consider donating to them.
- I learned about how to use a Raspberry Pi, Redis, and some engineering Rube Goldberg goodness to measure how much coffee is left in the pot from James Alexander[2].
- There were some amazing lightning talks on Saturday evening, made even more amazing by the fact that the projector didn't work for half of them and *they went on anyway* (more on this later).
- Stephanie Slattery[3] gave an *incredible* talk on accessibility and really inspired me to ensure that everything I do is as accessible as it can be. It's good for both ethical and financial reasons – how often is it that incentives align that well? Let's seize the opportunity and make the world better by making our tech improve the lives for all our users, instead of excluding a fifth of them.
- Andrew Wolfe[4] gave a great and humorous talk where he detailed how he built out the software for BrokerSavant, the challenges faced in scaling a machine learning pipeline (and some solutions!), and ironically technical difficulties started on a slide about unexpected technical difficulties.
- Our general counsel at DACA Time[5] attended a few talks and got visibly animated and excited about coding, which just, in so many ways, fills me with joy. Law and code aren't *that* different and are probably equally opaque to most people. Why not use one to solve the other?
- Katie Cunningham[6] discussed the ways in which technical inter-

---

[1] https://osmihelp.org/
[2] https://www.linkedin.com/in/yanigisawa/
[3] https://twitter.com/sublimemarch
[4] https://twitter.com/andrewwwolfe
[5] https://dacatime.com
[6] https://twitter.com/kcunning

views are often done *very poorly* and some of the ways you can fix it (often by just not doing things; for example, just don't whiteboard, it doesn't actually give you the info you think it gives you).

- Thanks to my medication[7], for the first time in my life, I was able to go up to two different speakers and initiate conversations with them. I was also able to initiate conversations with multiple audience members when I identified shared connections between us. This seems like a normal thing to be able to do, but for most of my life, I thought it was normal to just have crippling fear of talking to people, so I never initiated conversations with anyone else. (Again, consider donating to OSMI[8].)
- I had to miss a really good talk on mentoring because I had not eaten all day and did not want to pass out during my lightning talk later. The good news is that it was recorded, so I will still get to see it later!
- Lightning talks happened! As mentioned earlier, there were technical difficulties before but they were resolved. So I figured that I would be golden for my own, right? I was wrong. My laptop (running Ubuntu, so, you know) did not play nice with the projector, even with an audience member's adapter. What did I do? The only natural thing: describe and act out the GIFs I wanted to use. It was okay!

I had a great time overall, and I can't wait to post links to some of the videos of these great talks. (And hopefully there will be an embarrassing video in which I act out some cute GIFs.)

---

[7]https://en.wikipedia.org/wiki/Escitalopram
[8]https://osmihelp.org

46 On estimates, time, and evidence

Originally published on 2017.08.07.

Here's an exchange that's pretty common:

> "How long will that take?" "A few days."

I run into this all the time with clients - they have real business needs to know how long something will take and what the risks are with any given project. So, we are asked to give estimates of how long tasks will take. Whether in time (2 days) or points (3 points) later used to measure team velocity, these are ultimately an implicit agreement of roughly how long a task will take.

Much has been written about software estimation techniques. It is alarming how few citations are in these articles, however, given that the claims they make are verifiable – "X technique is more accurate than Y technique". For a field that claims to be quantitative and data-driven, we use alarmingly little data in our decisions of which tools and techniques to use (ironically, this claim is not one I have data to back up).

While reading "The Senior Software Engineer"[1], I came across a claim within it: when you are estimating a task, you will be more accurate if you estimate 1 day's worth of work than 1 week's worth of work, and more accurate if you estimate 1 week's worth of work than 1 month's worth of work. On the face of it, this seems like a very useful result if it is true - unfortunately, no citation was given. So, let's dig in.

Here is the claim: given two tasks T1 and T2, an estimate will be more accurate if it is for a shorter span of time. There are two subparts to this:

- What does it mean for an estimate to be accurate?
- Which way of doing estimates is the most accurate?

---

[1] http://theseniorsoftwareengineer.com/

46.1 What is accuracy?

Let's assume we have a task T and for that task, we have the estimated time, TE, and the actual time taken, TA. Two possible measures of error come to mind: raw time difference, and percent difference.

Raw time difference: Error = |TA - TE|

Percent difference: Error = |TA - TE| / TE

In the real world, raw time difference is going to be the most noticeable error, so it may influence how we perceive the accuracy of estimation techniques. On the other hand, percent difference is a more fair comparison, since it allows us to compare wildly different timescales: a raw difference of *one day* is clearly very significant if the initial estimate was *one hour*, whereas it is relatively inconsequential if the initial estimate was *one year*. For the purposes of this article, I will use percent difference when I refer to error, although it is helpful to keep in mind the raw time difference measure as it influences how we perceive accuracy and thus how we perceive different estimation techniques.

46.2 How We Perceive Time Matters

There are three possible worlds, and our goal is to determine which is the actual world and which are the counterfactual worlds. These world are ones in which: 1. estimates are likely to be more accurate if they are for a *shorter* time 2. estimates are likely to be more accurate if they are for a *longer* time 3. length of tasks has no impact on the accuracy of estimates

Many of my coworkers have espoused a belief in world 1, as did "The Senior Software Engineer", so I suspect that that's the industry consensus.

Let's run through some scenarios to see what these worlds would look like, if they were the actual world. For all the worlds, we will assume that the shorter task, T1, is estimated at 1 week and the longer task, T2, is estimated at 1 month.

In World 1, the shorter estimate is more likely to be accurate. For the sake of arbitrary numbers, let's say that T1 ends up having 10% error and T2 ends up having 30% error. In this situation, T1's raw time difference would be 0.5 days, and T2's would be 6 days (assuming 20 working days / month, and 5 working days / week). Ouch, that's a lot of slip!

In World 2, the longer estimate is more likely to be accurate, so we'll say that T1 ends up having 30% error and T2 ends up having 10% error.

T1's raw time difference would thus be 1.5 days, and T2's raw time difference would be 2 days. That's still a lot of slip, but the gap has narrowed significantly.

In World 3, the estimates are equally likely to be accurate, so we'll go in the middle and use 20% error for each. In this world, T1's raw time difference would be 1 day, and T2's raw time difference would be 4 days.

World	Error (1 week)	Slip (1 week)	Error (1 month)	Slip (1 month)
1	10%	0.5 days	30%	6 days
2	30%	1.5 days	10%	2 days
3	20%	1 day	20%	4 days

Table 1: error and slip (raw time difference) in all three possible worlds.

Note that in all three possible worlds, the raw time difference in a 1 month estimate exceeds the raw time difference of a 1 week estimate, and in worlds 1 and 3, the differences are significant to the point where other confounding factors will probably play a larger role in the total amount of slip than just which of these worlds you are in.

The point of this exercise is not to show you that we are living in world 1 or world 2 or world 3. The point is to show you that in all possible worlds, it is likely that the slip from a 1 week estimate will be smaller than the slip from a 1 month estimate and that this has *absolutely nothing* to do with whether or not shorter estimates are more *accurate* than longer estimates.

This colors our overall perception of whether or not shorter estimates are more accurate than others. Managers and engineers alike will remember a slip of 4 days or 6 days as "about a week", and they'll remember a slip of 0.5 days or 1 day as "a little behind schedule", so at the end of the day world 1 and world 3 both seem like they will favor the mental model that shorter estimates are more accurate, even though that is not true in world 3! The fact that these two very different worlds are difficult to tell apart from "on the ground" should alarm us.

46.3 Let's Use Evidence

Because our perception can be heavily biased by a lot of factors - as shown above, but also by what we want to be true - we should lean on evidence and scientific studies to determine what is actually true.

It turns out that even this simple question (are shorter or longer estimates more accurate?) does not readily turn up in the academic literature. This is likely due to my inexperience with searching academic literature (I completed a grand total of one semester of a doctoral program). That inexperience is likely shared among my fellow engineers, and my peers may also not have readily available access to academic literature (fortunately, my undergrad university lets us keep library access for a long time after graduation). The combination of lack of exposure and lack of access to journals makes it fairly unsurprising that our books and blog posts do not reference the literature. It does not make it any less disappointing.

In general, studies show[2] that we are overly optimistic in our time estimation, such that in complicated tasks, we will be more likely to hit a schedule overrun than in less complicated tasks (and longer tasks are probably more complicated than shorter tasks). Here's a quote from the survey paper:

> In sum, the results suggest that bottom-up-based estimates only lead to improved estimation accuracy if the uncertainty of the whole task is high, i.e., the task is too complex to estimate as a whole, and, the decomposition structure activates relevant knowledge only. The validity of these two conditions is, typically, not possible know in advance and applying both top-down and bottom-up estimation processes, therefore, reduces the risk of highly inaccurate estimates.

Decomposing tasks into smaller units of time is helpful when the uncertainty of the task's duration is high, and looking at the task holistically is helpful when the uncertainty of the task's duration is low, and we can't know which it is until we get through the task, so let's do both!

This matches my intuition. Some large tasks that are straightforward are easy to estimate accurately even though they take a long time: for example, I could tell you with great accuracy how long it would take me to drive my car from my home in Columbus to my inlaws' place in Philadelphia, even though I don't know exactly where we will stop in the middle or for exactly how long. Some small tasks are not straightforward to estimate accurately: it may take three seconds to get my cat into her carrier, but if she's in a feisty mood, it may take as long as ten minutes, or longer.

I still haven't found an evidence-based answer to the question of whether or not, in general, shorter tasks are more accurately estimated than longer tasks. There are a lot of confounding factors, like how you do estimates in general (which will likely change when you go to estimate a larger project!). I'm not even sure that it's an important question to answer, because the actual accuracy of the estimate is probably not the largest driving factor in deciding how you approach doing estimates.

---

[2] http://simula.no/publications/review-studies-expert-estimation-software

What *is* important is making sure that we have data to back up our claims when we assert that certain methodologies are better than others. These are testable claims - let's test them.

Here are some testable claims that I would like to see answers to (note: I haven't actually searched for answers to these; but I *have* seen many people, including myself, assert these are true or false without any evidence, just anecdotes):

- Functional programming makes it easier to write parallel programs
- Functional programming results in less buggy code
- Agile development increases development speed
- Shorter estimates are more accurate than longer estimates
- Open offices are better for productivity/collaboration than individual offices or team offices
- Type-checked languages have fewer production bugs than dynamically typed languages

These are just a few of the claims that people make, without evidence, which are testable.

47 The bittersweet end of a year of independence

Originally published on 2017.09.02.

Just over a year ago, I left the startup[1] I was working for and started my own business[2]. My intention was to do freelance work ("consulting", to all my clients) until I was able to launch my first product, and then shift into being a product company. My ambitions and confidence were very high. In this last year, I have accomplished a great deal and have a lot of pride in the work I did, as well as what I have learned. Nothing took the path I expected it to, but I wouldn't change that at all. With that in mind, sadly, I am winding down my consulting work and taking on a new full-time job. I'll explain why at the end, but first I want to share some a little bit about what I have experienced in the last year and why it was valuable.

47.1 Why I left my job

It's valid to ask why I would even bother starting a business. It is a much harder path than getting another full-time job, with more stress, and with more risk. There are, however, some simple and clear reasons why I left my job. There were three primary reasons I left: I wanted to work on things with more autonomy; I wanted to control my work environment more; and I wanted to make more money.

Autonomy: one thing that is important to me is owning the ideas/products I'm working on and developing them holistically, with a stake in the results. At the end of the day, I am more motivated if I am working on something very important to me, and it leads to greater results. I also thought that if I were working on my own I would develop better skills since I would need them and could not lean on anyone else for that; this was later proven correct, since working on my own projects led me to learn front-end development with an urgency I could not have previously imagined.

---

[1]https://crosschx.com/
[2]http://caturra.io/

Environment: my previous company had an open office, with lots of exposed concrete, metal, and wood. Needless to say, it was a very loud environment. I've learned that I am simply not productive in that environment. Among other issues, I have misophonia[3] (it is at its worst when I am tired or stressed), a peanut allergy, depression, and anxiety. These all made open offices very difficult for me to work in, and controlling my own environment and working remote has led me to being far happier and far more productive than when I worked in a physical office. Everyone is different; for me, controlling my environment has made a world of difference in ways I could not imagine.

Money: It was no secret at my last company that we were underpaid. My manager told me as much. This wasn't enough to make me want to leave on my own, but combined with a desire for more autonomy and for a work environment that worked better for me, it definitely increased my motivation to leave.

At the time I left, my reasons for leaving were not quite as clear to me. I had some reasons and I had the story I told as I left. It was not a lie: I did want to leave to work on my own products. It just took this last year for me to fully realize *why* I wanted to leave to work on my own products. At the end of the day, that office environment was not a good fit for me, and in the absence of a good fit, a lot of other small issues become big issues.

So with that, I left with grand ideas of a few products I could make, and had a few clients lined up to keep the money rolling in until my products were launched.

47.2 How I spent the last year

When I left my job, I gave myself a plan. I would spend about half of my week working for clients, and I would spend the other half of my week learning new skills and working on my products. None of my product ideas worked out, because I did not have the skills to do front-end development when I started trying to make some web-app products. I did learn a lot, and actually worked on some very cool client projects.

For one major property management software company, I rewrote their ETL pipeline using some big data tools and techniques so that it could scale and could run two orders of magnitude faster. This client's work was boring in some ways, but I'm indebted to my friend who introduced me to the team (I owe you a coffee, if you're reading this) because landing this client gave me the ability to quit my job.

---

[3] https://en.wikipedia.org/wiki/Misophonia

I also worked with the world's biggest bureaucracy[4] to modernize some of their old data systems and make it so that some really important data is more accessible, thereby enabling the internal teams to save real lives. This project was awesome in many ways, because it's rare to work on a project that has such a clear line to lives saved. It was also frustrating in some ways, which hopefully I'll be able to write another post about.

Along the way I had some various small clients, who I consulted for on data engineering related topics, built small web-apps, etc. These were nothing to write home about, but they did give me a lot of insight into business and the value my code can add (or the lack thereof).

In April, I also co-founded DACA Time[5], developed the prototype, and built up a small team of volunteers to help me with some of the development tasks. This would not have been possible if I had been traditionally employed, since I was spending 20+ hours a week on this at some points.

47.3 Flexibility saved my life and my career

I started my business so that I could have the flexibility to develop products while still paying my bills, but flexibility turned out being valuable to me for many more reasons than just that, in ways I could not have predicted.

First and foremost, I believe that having flexibility saved my life and my career. In February, I was diagnosed with depression and anxiety. It was bad at that point: I had attempted to harm myself; I was only functional for 20-24 hours per week (I could work, then I would shut down); I had no interest in doing anything and was considering quitting tech entirely; and I spent probably half my time curled up and crying. Let me repeat that: my friends and clients had no idea that anything was wrong, but I was barely holding it together during work and was seriously considering doing permanent damage to myself or quitting my line of work entirely.

I believe that if I had had a normal job, I would have not been able to hold it together even that long. That may or may not have been better for me, but I do know that having flexibility made it a lot easier for me to get to a doctor to seek treatment, and it made it a lot easier to take time off for mental health.

This flexibility is also what led to me attending GiveBackHack[6] and co-founding DACA Time[7], which both showed me how much I can do as a

---

[4] http://www.un.org/en/index.html
[5] https://dacatime.com/
[6] http://givebackhack.com/
[7] https://dacatime.com/

software engineer, and reinvigorated my passion for software engineering, product design, and making a damn difference in the world.

47.4 Consulting taught me a lot

During the course of the last year, I expected to learn a lot, and I did - but not the things I expected to learn. I expected to dive deep into machine learning, AI, and data engineering, and become a world-class expert in my narrow niche. It turns out, running a business actually doesn't teach you advanced mathematics, but does teach you some other practical things - who knew?

Being a consultant let me see how businesses worked on the business-end of things, rather than just the development side. I learned more about how my work directly impacts revenue, which is a lesson I will carry close to my heart through the rest of my work.

I was also better able to determine my market value. When you're on your own, every client you get is a chance to re-negotiate your pay, so you can try over and over and eventually have a really good idea of your market value. I still don't know what my consulting rates should have been, but my clients were *way* too happy with the price they paid for those rates to have been close to what the market would bear.

The importance of networking and communication was also made really clear, since all my clients (literally every single one) came from my network. Focusing on communicating complex technical details to non-technical clients or less-technical folks became very very important, and made me realize how much value can be added just through clear communication; or how much value can be lost when the details are not communicated clearly. If no one knows who you are, what you can do, or what you did for them, then you cannot deliver them any value.

47.5 Go forth and start a business

If you are at all on the fence about starting your own business, you should do it. You will learn a lot about yourself, about the business world, and possibly about software development, and you will come away from it a much stronger contributor than if you just remained a normal software developer. You're better off taking the plunge and finding out that you don't like it, with some great stories to tell, rather than wondering if you could have or should have done it.

If you are considering this and want to talk about how to get started, reach out to me and we can set up a coffee or a chat sometime.

47.6 Why I'm winding down my business

Self-employment has treated me really well, and I am in a much better position than I was a year ago in terms of happiness, fulfillment, and mental health. So why am I leaving self-employment behind to take a full-time job again?

Well, there are a few reasons:

- My wife and I are both self-employed. This creates are few challenges. Good insurance is super expensive, and my mental health treatment this winter/spring made me painfully aware of how expensive ultrasounds are. Additionally, banks are unfortunately *not* very fond of lending money to two self-employed people, especially since I do not have a long history of it.

- I really really miss being part of a real team. When you're a consultant, you just have a very different relationship with everyone on a team than if you are a member of that team, and it's very isolating. When combined with being 100% remote and having less human contact, this can be challenging. I want to be part of a team again so we can rally together to do great things, so we can lean on each other, so we can be *friends* instead of being clients/consultants.

- Consulting just isn't making me happy. My skill is as an individual contributor, not at running a business or being a manager. Running my own business required me to manage a lot of aspects of the software development process that I'm not good at, and it required me to manage a lot about my business that was very inefficient for me. (Next time around, and I promise there will be a next time, my wife is going to help with the business side of things, and I will outsource as much of the rest of these tasks as I can.)

So on that note, I'm really happy and sad at the same time to say that I'm going to stop working as a consultant and will be moving back into a full-time job. Some of my friends know what company I'm joining, but it isn't public until after I've officially started (if you're curious, watch my LinkedIn profile). This is really bittersweet for me. There are so many advantages and good things about being with a company, but it comes with a certain loss of freedom and autonomy as well (and a loss of time to put towards DACA Time). I'm really confident that the team I'm joining is a great one, composed of great people, so I will be able to retain a lot of the flexibility which I have thrived with (otherwise, I wouldn't do this), but it remains a bittersweet end to a year of independence.

48 How I work remotely

Originally published on 2018.06.02.

I've been working remote since September 2016. There are a lot of engineers who have worked remote longer than I have; there are others who have more insight into how they work than I do; and there are plenty of people who simply don't work in the same way I do. My intention in this post is to share how I work, the reasons why I work that way, and what I think others should try while finding the process that works best for them and their teams.

48.1 Remote Work, Round 1

In September 2016, I joined a team as a remote engineer for the first time. I had just recently left a full-time traditional software engineering job to pursue my own company: I was splitting my time between 50% contract work / consulting and 50% personal projects with the goal of creating a startup. (I managed that, and co-founded a startup non-profit[1] which aims to make the barrier for immigration whether you *qualify*, not whether you can *navigate bureaucracy*.)

This was a learning experience for me in more ways than I was prepared for. This year of independence taught be a lot about time management, prioritization, the sheer difficulty of starting and running a business (let alone two at the same time). It also taught me a lot about how to work effectively. I'm going to focus on that: the mistakes I made and lessons I learned which increased my productivity and happiness as an engineer.

In many ways, it's easier to identify what not to do, rather than what to do, so I'll start there.

My first real remote-work experience was as the solitary remote engineer on an otherwise colocated team. I put my head down and **focused on pure productivity, ignoring personal interactions**. This worked well in some respects, because the code I wrote was really good code and achieved its purpose. However, it failed to recognize one important aspect of that work: yes, I was a remote engineer... *on a team*. I never

---

[1] https://www.dacatime.com

built cohesion with the rest of the team, which led to some suboptimal outcomes.

I also communicated **at a level I thought was appropriate, and avoided over-communicating**. What I have found since then is that it is almost impossible to over-communicate (I would say that it *is* impossible, but I tend to avoid absolutes). More on over-communication later; for now, suffice to say that a lack of communication leads to decreased visibility, clarity, and rapport.

For another client, our project ended up having a mismatch between delivery and expectations for one team member. We expected a certain outcome, he expected a different outcome, and at the end of the day, the stakeholders were unhappy with what we delivered. This, too, was a result of **not checking in with the team and building a rapport**. If we had had more frequent check-ins as a team and had more rapport built-up, then it would have been much easier to both detect the problem and to course-correct for it.

A lot of these mistakes can be boiled down to highlight what it is important to value:

- Frequent clear communication
- Team cohesion and rapport

My observation is that engineers tend to be singularly focused on *shipping* and less focused on the other aspects, so deliberate attention toward these helps avoid these kinds of mistakes. A team of remote engineers is still a *team*, and the team aspects of the problems will not be solved unless you, dear reader, approach them with intention.

48.2 Leveling Up

In July 2017, I joined Remesh[2] as the third engineer, and the first remote employee. I knew I had to approach remote work with more intention to win the trust of the team–not just to protect myself and my job, but also to avoid giving a negative impression of remote work in general. Since then, we've hired more remote engineers and I'm still employed, so I would say it has gone well!

In spite of the mistakes I made in remote work previously, I was still an effective engineer. With this new job, I wanted to make sure I was not just effective, but could set others up for success as well, as the team grew. To learn more and refine my approach, I read Cal Newport's book

---

[2]https://remesh.ai

Deep Work[3], Julia Evans' excellent remote work blog post[4], and countless posts on StackOverflow, Reddit, and HackerNews about how to do this effectively. I ended up with an approach that works very well for me and which may be useful for others.

What I have found is that the most important thing to work on as a remote engineer is **communication**, and your **working style** is also key to your individual and team success.

48.2.1 Communication

Communication is where a lot of teams break down, especially teams which are a hybrid of remote and colocated engineers (one of the most challenging team architectures, in my opinion). Communication takes active effort to learn and is certainly not taught in computer science curriculums, which is part of why you primarily see senior engineers working remote: junior engineers need more active, personal, face-to-face interaction to develop their craft. It is doable if you put some effort and intention into it, and here are some maxims which I've found work well.

48.2.1.1 Maxims for the Remote Engineer

- **Always overcommunicate.** You can't actually achieve this, so trying to get there is a good way to ensure that lines of communication stay open and everyone knows what you're working on, how you're doing, what you're struggling with, etc. and views you as more than a couple of comments on a GitHub issue.
- **Let people know when you're in or out.** There's a tendency for colocated people to have no idea when remote people are on or off, because they can't see you, which leads to assuming that you're either always reachable or always unreachable (frustrating either way). Saying when you come online or are leaving for the day helps set expectations and establish a rhythm, just like when you see your buddies at the coffeepot in the office in the morning, and walking out at 6pm. Similarly, say when you step out for a moment to go for a walk or head to the coffeeshop, as well.
- **Ensure that you are reachable for emergencies.** This usually just means: put your SMS number in your Slack profile so that if we need to find you, we can. Details vary by company. If there isn't a good way to discover your coworkers' contact info, suggest a system for it (can be as simple as a spreadsheet of phone numbers and timezones).

---

[3] http://calnewport.com/books/deep-work/
[4] https://jvns.ca/blog/2018/02/18/working-remotely--4-years-in/

- **Uninstall Slack from your phone.** I'll wait, do it right now. Get rid of email while you're at it. The reason for this: as a remote worker, boundaries between work and life are already blurred, so it takes extra intention and effort to actually establish separation between work and life, which will boost your productivity and make you happier.
- **Practice clear and concise written English.** We're programmers, and a lot of us were probably (unfortunately) in that group that made fun of English majors. Turns out, though, writing well is really damn important and it benefits everyone to run a spellchecker, proofread for grammar, and make sure your messages/emails are well-written and structured logically. It only takes a few minutes to do this, and it will save you and your coworkers a lot of time by making things clear the first time, instead of requiring a back-and-forth. Also, try to write with more formal English, not how you text your friends: it will be clearer to more people, and it will project more professionalism.
- **(Controversial) Use tons of emoji** 😁. It's hard to tell someone's tone without body language. Emoji can help convey tone and at least make it clear if you intend something to be funny or not.
- **Schedule unstructured time with coworkers.** You know those water cooler conversations you have in a real office? You know how you bond over lunch? We don't have that, so you have to put in deliberate effort to construct those same interactions. I've scheduled a bunch of biweekly touchbase meetings with my peers and have gotten a lot of value out of them (including a discussion which led directly to me writing this blog post). These meetings spark interactions which wouldn't happen otherwise, and they're valuable precisely because they have no plan and no agenda. I suggest scheduling these with the people you work closely with, those you do similar work to but aren't close with, and other people across your organization. I also have been loving our coffee buddy[5] app which pairs random people every two weeks; now I've talked to a lot of people on the business side of the organization and gotten unique insights into our product.
- **Talk about personal things with coworkers.** It's important to develop bonds with your coworkers. I had no idea that my coworker Dan is as into coffee as I am until it came up in conversation in our NYC headquarters, but now we have something to break the ice and chit-chat about, leading to higher team cohesion, happiness, and productivity.
- **Schedule deep work time.** One of the key benefits of remote work is the ability to easily enter into deep work. However, when you do this, manage expectations and let your coworkers know through

---

[5]https://slack.com/apps/A11MJ51SR-donut

your Slack status, calendar events, etc. that you are doing deep work and are not reachable. This manages expectations and leads to less frustration from them because it's clear why you're not responding, and less frustration for you because they're less likely to keep pushing to punch through and notify you.

- **Use calls whenever it makes sense.** Even though a lot of remote work is asynchronous, a phone call is often the most efficient way to quickly hash something out and unblock someone. It's less frustrating to talk for 5 minutes than to have 20 emails back and forth. Don't call someone if they're in deep work mode, but if you're actively chatting with someone, consider if a call would be better than using text.

48.2.1.2 Maxims for the Colocated Engineer

If you're on a team with remote engineers, it is helpful to intentionally work to enable their inclusion. Here are some maxims which I've found are helpful to follow to include your entire team, not just your colocated team.

- **Don't treat colocated as the default.** Even if your team is 90% colocated and 10% remote, if you refer to colocated engineers as "the engineers" and the remote engineers as "the remote engineers", then you are other-ing the remote team members. It is helpful to simply acknowledge that these are two categories of your employees, and make neither the default in your language.
- **Default to text first.** If you're discussing something and you *can* do it via Slack or email, do it via Slack or email. That way everyone can participate, not just other colocated engineers.
- **Let other people know when you're in or out.** Just like you can't see when remote people sign on, it's super helpful to say on Slack when you arrive in the morning, are going out for a coffee, or are heading home for the evening, so that remote people know if you're reachable or not.
- **Ensure that you are reachable for emergencies.** Exactly the same as above.
- **Practice clear and conscise written English.** Exactly the same as above.
- **For one-off meetings, mention them on Slack.** Your colocated coworkers can overhear an interesting meeting and chime in, but your remote coworkers cannot. So if you mention something like "Hey, I'm talking with @AwesomeEngineer about CoolTopic right now," then people can respond with "Oh hey, I had thoughts on that, can you loop me in?" or "That sounds interesting, mind if I

eavesdrop?" This will lead to more insights and more knowledge transfer among the team.

- **In meetings, use raised hands or a passed object to get input.** If you rely on body language to determine who speaks next in a conversation, colocated coworkers will dominate the conversation because remote workers cannot express much body language on a call (and the call's speakers are usually quieter than a colocated person can be). If you rely on raising hands or passing an object to pass the metaphorical mic, it is much easier to see if a remote person has something to add and loop them in.

These maxims are what I've found to work for me, and are not universal laws. If you have something else you think should be included (or something which shouldn't be), email me@ntietz.com and I'd love to have a conversation about it!

48.2.2 Working Style

Everyone has a different working style: morning people who get up early (hi), night owls who work late, some people work super long hours, some of us have strong work-life boundaries. Here is how I've set up my working style. I think these transfer well to colocated practices, as well, and I'd encourage trying them out.

- **Set a standard schedule.** Yes, you'll deviate sometimes, but having a standard schedule achieves two things: it gives predictability/reliability to those on your team if they need to reach you; and it makes it so that you can rest and recharge when you're outside of work. Remote engineers don't have as strong of a physical separation between work and home, so a temporal separation is very helpful.
- **Maintain a physical separation.** It's very hard to dissociate work from the place where you work, so set aside some of your space for *just work* and don't do anything there unless you're working. This also makes resting and recharging easier when you're offline. I'm fortunate to have a separate room of my house which is my office and is used just for work, but this space could be as simple as a desk in your living room that's used for work and only for work.
- **Visit colocated engineers occasionally.** It's important to get this face-time, especially with new team members who you have had less interaction with and with junior team members who benefit from more mentoring. And make sure you communicate your travel plans loudly and often so that everyone is aware of where you're going to be and when you will be there.

- **Make a daily agenda.** I write out all my priorities for the day and then try to fill my day (approximately 9 AM to 6 PM) in 30-minute chunks with what I am doing and when I am doing it. I rarely stick strictly to this schedule, and I believe the core benefit is going through the daily exercise of prioritization and estimating how much I can actually achieve; it keeps excess optimism in check. (If I don't do this, I tend to work longer hours and still get less done, because I feel overwhelmed and pressured.)

48.3 Conclusion

If you take away nothing else, take away this: approach your working style and your communication style with intention and iterate on it until you've found something that works well for your team and yourself. I've found an approach here which I think is very good and works really well for me and our team, but there is no one-size-fits-all solution.

49 Topologies of remote teams

Originally published on 2018.08.23.

When you're building or scaling a software engineering team, you naturally run into a choice at some point: will we all be in the same office, or will we do this "remote work" thing? There are a lot of factors that go into whether or not remote work will work for your team, like if you know how to work remote[1]. Another consideration, to make it *more* complicated, is which form of remote work you want to consider.

There are four different "topologies" of remote work that I've observed:

1. The Linux model: fully remote, fully asynchronous
2. The Basecamp model: fully remote, somewhat synchronous
3. The hybrid model: half remote, half colocated, fully synchronous
4. The traditional model: colocated team, possibly with some remote team members

I've been on three of these types of teams, and I've seen the other quite a bit. Let's take a deeper dive into each of them, and then talk about how to make a decision at the end.

49.1 The Linux model: fully remote, fully asynchronous

This model is common in open source software projects, due to practical concerns: people work on it at odd hours and cannot be expected to be on chat all at the same time.

You can get a lot of great work done this way. This blog post was written using software that was mostly created this way. The Linux kernel certainly was, at any rate: Linus Torvalds uploaded the source code and sent out some emails on a mailing list, and then other programmers were able to send patches in. As far as I know, the Linux kernel developers don't hop onto Slack to talk all day and to have video calls, so most of their communication is through asynchronous means like email.

This model will work for you when the people on your team work very well with high degrees of autonomy. Since it's asynchronous, they have

---

[1]https://ntietz.com/2018/06/02/remote.html

to be able to do this, or they will run into periods of indecision and stall out.

This is the model which I haven't lived first-hand. My evenings aren't filled with open source contributions (I'd rather spend the time cooking or reading a good book or writing *English*). I haven't seen this at a lot of companies, although there are some companies where it's debatable if they fall in this category or the next one.

Personally, I find this one suboptimal; it's really nice to have a couple of hours each day where you overlap with the coworkers you're working closely with so you can bounce ideas off them directly, whether it's for debugging or for designing a new feature or for solving a gnarly architecture problem. But it can work and it emphasizes deep work, so there's a big benefit there.

49.2 The Basecamp model: fully remote, somewhat synchronous

This model is probably the prototypical commercial fully remote model. This is how I'd categorize companies like Basecamp, GitHub, and others. They're almost entirely remote but they tend to have at least a few hours of overlap of timezones between people who are working closely together to allow for those immensely valuable interactions where you put two people together but get more than two times better solutions as a result.

This model works well when people on your team can work with high degrees of autonomy but they don't have to be quite as autonomous as when it's fully asynchronous, since you have some overlap to bounce ideas off of people and get input from them and talk about where you're going next.

The pitfall in this model is that some activities are simply harder. To the best of my knowledge, there is no great way to whiteboard together with remote employees, and that's a great technique for designing software. It's also difficult to pair program, and mentoring junior engineers just has higher friction when remote.

That said, this is an incredibly fun and productive way to work. I did this on a contracting project, and it was really great for the freedom of it, since you could be offline at any time as long as you were getting your work done. The biggest drawback, on our project, was that it was difficult to get a rhythm going due to all of us having different schedules (despite all living in the same timezone, we actually had few overlapping work hours), which just emphasizes the importance of overlapping work hours and also how hard schedules are, even for a handful of people.

49.3 The hybrid model: half remote, half colocated, fully synchronous

For pragmatic reasons, more and more companies are adopting this pattern. The company I work for, Remesh[2], does this: we have engineers at our HQ in NYC, but we actually have a slim majority of our engineers spread out across the US. We got into this model because the team had two engineers and needed to staff up, so they brought me on as the first remote engineer; it went well, so we gradually hired more remote engineers.

In general, this model is charcaterized by a very strong geographic presence in one location but with a large number of remote engineers. Because these teams have a lot of colocated engineers, they tend to emphasize having a lot of overlap in their days, prioritize synchronous communication, and have high team cohesion.

This configuration has a lot of benefits. Having a lot of colocated engineers makes it easier to build team cohesion. But by having so many engineers remote, you are forced to adopt remote-work patterns. The whole taem benefits from better documentation and more location indepependence. Not to mention, you also reap one of the biggest benefits of remote work: the gigantic pool of talent out there, since *most* of the talented engineers don't live where you live.

The biggest difficulty here is keeping cohesion between your colocated and remote team members. The colocated engineers will tend to form tighter bonds because they see each other every day. There are some ways around that, like having frequent meetups for the remote team members, but it's a risk factor you just have to be aware of and have to work to mitigate.

If you go this route, make sure that you actively engage the remote engineers, and consider forcing all the engineers to spend *some* time remote to build empathy and stronger habits on the team as a whole. Think of it as chaos engineering[3] for your team: if you randomly prevent people from working inside the office, you will *force* your team to document better, be remote friendly, and be more independent and autonomous.

---

[2] https://remesh.ai
[3] https://techcrunch.com/2018/02/04/the-rise-of-chaos-engineering/

49.4 The traditional model: colocated team with a few remote team members

This one is easy to identify from a distance: everyone is in one location except for a few loners who are remote. This usually happens in traditional work environments when something major changes: either an employee is going to move to another city and the company makes this work to keep them on; or they need to bring on talent for a specific skillset and they cannot find it locally.

I don't recommend this model. The benefits are minimal, and are just centered around a specific person that you want to be able to work with. But the drawbacks are huge. You will have a lot of difficulty integrating the remote person in, since your work patterns are all set up for colocated engineers. You will struggle to retain this remote employee for a long time, since you will in all likelihood alienate them or they will simply feel left out by being unable to participate in local team events. It can work with people who you have a really good rapport with or who are already very, *very* good at remote work, but it usually does not work well.

The only situation I'd recommend this for, as the employer or the employee, is for short-term contracts. For anything beyond that, if you want to embrace remote work, go all in and at least embrace the patterns that will make it work well, since those will benefit your colocated employees and you will broaden your talent pool.

49.5 Recommendations

Of these, the "traditional" model is to be avoided at nearly any cost since it is painful with few benefits, and the Linux model is ill-suited for most businesses since you sacrifice agility of decision making which is critical to launching good, fresh products.

So we're left with two realistic models: fully remote but mostly synchronous; or half remote and half colocated. If you execute either of these well, you reap tremendous benefits, so the difference comes down to how much you value colocation. I personally don't find it intensely valuable, so I'll always vote remote; others do find it valuable. One way to make this decision, to find out how much colocation is important for you, is to simply *try* to be fully remote. With your existing team, close the office for a month and see how you all get on; or if you're scaling up, require half the team to work from outside the office half the time, so you don't have to get that bigger office space just yet. It will be difficult at first, but it will tell you whether or not you truly need colocation and it will expose what you were getting from it - and what it was costing you. And then you will know which way you should go.

Personally, when I start a company, I'm going to go all-in on remote work from day one.

50 Even bad estimates are valuable if you use them right

Originally published on 2018.08.31.

Estimating software projects is hard, if not impossible. This seems likely to be fundamental to the work, because we're inventing new things and invention doesn't happen on a fixed schedule. And yet, many teams still estimate how long their tasks will take to finish. Why should you do this, if you can't do it accurately? You do it because it can help you reach your real goal of solving a problem as quickly as possible. But when you do it, you need to have really solid processes around estimating, or the estimates will be used and abused and can kill your team.

Let's establish a baseline first: what's an estimate? It's a measure of how long a piece of code is expected to take to complete. This includes the time you need to do non-code tasks, like reproduce a bug or model your data. This includes the time it takes to test your feature, to write automated tests, and to go through the code review and QA process, since those can lead to code changes. Simply put: it's the total amount of time that you expect any member of your team to invest in this change, in any way.

You can do these estimates a few different ways[1], such as with story points, t-shirt sizing, or time buckets. One important thing to do, regardless of which metric you use, is to think about and quantify your **uncertainty**: if you're highly uncertain of an issue's size, then you might want to timebox some investigation into the issue to reduce the uncertainty and de-risk it. These estimates, of any type, are useful to let you know when things are going off the rails[2]. Each sprint, you decide on what your team is trying to accomplish. During the sprint, you let everyone know what you're working on and what you're blocked by at a daily standup. That standup is generally the place for you to say "Hey, I'm working on feature X, but it's turning out to be a lot more complex than we thought; could anyone see if I'm missing something, or should we reduce scope on this?" Then your team can make an informed decision and you can

---

[1] https://producthabits.com/engineering-estimates/

[2] This is why I use clock time for my estimates, rather than story points or t-shirt sizes. When you're using them to adjust course mid-sprint, you need to be able to quickly tell if you're going off the rails. That's much harder with t-shirt sizes, since you need to convert from the size to clock time and then compare your progress - and the sizes don't correspond to exact clock times, anyway!

either change course to reduce scope, remove some blockers, or charge ahead as planned and accept that this task is more complex than you anticipated (it happens!). But without these estimates, you're flying blind, and you'll just **always** charge ahead, missing opportunities to reduce scope or collaborate more with your team members.

With estimates, you also are forced to think through things at the beginning. You switch from fast, instinctive thinking into slow, deliberate thinking so you find the true complexity of issues rather than assuming their surface level simplicity is accurate. This is incredibly helpful in reaching where you want to go, because it leads you to focus on creating the shortest path to a solution which you can test with users. If creating a login page is super complicated, well, do you **need** the login page to test your app with real humans? Or can you hack it, using an identity-as-a-service provider or even using **no** login for hands-on user trials?

Doing estimates does have drawbacks, however. You need to have buy-in from everyone your team interfaces with, as well, or you risk Deadline Driven Development. If you have solid estimates and the business team gets their hands on them - without explanations from you - you can expect that these features will be promised on some form of timeline. So, you must **explain** to your stakeholders beforehand that these estimates are only for course correction during the development process, and they're separate from estimates you will give of when features will be done overall. If this isn't done, you can lose trust on your team, you will lose trust of the people outside of your team, and morale can drop precipitously.

The other main drawback is simply that providing estimates takes time, which is time you could spend just writing code instead. If you never use the estimates to adjust what you are working on, then putting in the time to do estimates is a pure waste. However, if you do put in the time to do estimates, you will spend less time coding - but because your team will be able to respond to things immediately, you will still reach your objective more quickly.

An ideal scenario for estimating and using them well looks like this:

- You are using two-week sprints, within the context of a larger goal (solve problem X)
- You have daily standups which everyone on your team attends
- At the beginning of each sprint, you plan what everyone is working on and estimate it to ensure that it's an appropriate amount of work for one sprint (you may also add "background tasks" to fill time when people are blocked)
- Every day, you run standups to see what's at risk of going off the rails and what's blocking progress so the team can get out in front of it

- Whenever things look like they might go off the rails, you reassess and adjust course: shrink scope, expand estimate, or remove blockers
- Throughout the process, everyone outside of the team either cannot see your estimates or understands that they are **not** deadlines or promises

So go forth and try doing estimates, and see how it goes! It's challenging, but you can improve at it quickly, and the benefits are really great for doing them, especially in a team environment. You will quickly find that you can anticipate issues more quickly and that you think about risks earlier in the project. Just don't let your business team make promises based on them!

51 Don't Disrupt Things; Fix Them

Originally published on 2018.09.07.

People talk about disrupting industries when those industries appear to be in a stable but inefficient state. For example, the taxicab industry: there was little innovation going on in it, and it was stable, but it seemed like it was far from ideal. Along came Uber, intent to disrupt the industry - and disrupt it they did.

Uber has a culture of ignoring laws around the world when it's convenient to do so, when it helps them earn a buck and disrupt the existing industry. As a result, we have an app that millions of people love and use regularly, and which provides income to millions of drivers. This seems pretty good, in this framing: we got something we all like using, and people are earning money from it.

However, there is a darker framing to it. If Uber disrupted something, what did it disrupt? It disrupted the livelihoods of the millions of taxicab drivers.

The drivers who existed in the old, stable-state system were obeying the laws, in general. They played by the rules, even when it was expensive for them to do so. Taxi medallions in New York City cost over $1 million at their peak, and drivers or taxi company owners had to invest lots of money in this resource which appeared scarce. Now, those medallions are often valued under $200,000, having lost over 80% of their value. Uber has gained value by breaking the law (creating unregulated taxis) and law-abiding businesses have lost out by playing by the rules. The old rules did not make sense in many instances, as there was clearly artificial scarcity at play here, but that does not change the fact that the ethical businesses lost by being ethical.

Similarly, Airbnb has disrupted the hotel industry. The loser here? Big hotel chains which abide by regulations and pay their local hospitality taxes. And consumers, who are now staying in unregulated, potentially unsafe hotels rather than staying in hotels which are regulated by their local governments. Again, you can argue against paying the taxes and against the regulation, but what you can't argue with is this: the businesses who played by the rules lost, and the players who ignored laws and rules came out ahead financially.

Clearly, there is something broken here. It should not be a viable business model to ignore and violate local rules and regulations and then just pay fines down the road, because the economic impact to many is so great. The focus of these businesses should not be *disrupting*, but *fixing*. If the taxi industry is broken, let's fix it! If the hotel/hospitality industry is broken, let's fix it! But consider the side effects in the process. Consider who you're putting out of business and what will happen to *their* livelihood if you do disrupt their life.

Disrupting things is not *inherently* valuable for society. In fact, while a disruption will push you out of a steady state, you have absolutely no guarantee that you will be in a better position when you get to the new steady state. You could leave society a better place - but you could also make it actively worse.

Think about that next time you set out to solve a problem: instead of disrupting an industry, let's solve a problem and consider the wider impacts.

52 Distractions Cause Bad Code

Originally published on 2018.09.14.

We are barraged by constant distractions, and they are degrading the quality of our work. Our digital society now is set up to allow us to focus for mere minutes at a time, since we are in an attention economy and the sole objective of companies is to capture more of our time. Facebook, Google, and Snapchat are all incentivized to get us to look at our phones many times a day.

Distractions permeate everything, even at work. GitHub has notifications for so many things that if you have work and personal projects on the same account, you will get unrelated notifications all the time. Our employers set us up with ping-pong tables, open offices, and Slack, the open-office of chat tools.

With all these distractions surrounding us and with all these notifications, we are expected to get deep work done. Personally, I cannot. And you cannot, either.

When I'm highly distracted, I'm prevented from entering flow. To my core, I'm a maker. I get such a thrill from making things that are usable and useful, and these distractions cut through that in a way that makes it impossible to have a productive, fulfilling day.

Flow is important if you want to get anything meaningful done. Context switching takes a lot of effort and time and you can only do it so many times in a day. If you are constantly distracted, you will never enter flow and you will never have great, innovative ideas.

If you never concentrate and go deep, you will produce bad work: you will produce bugs, and fail to debug them; you will create security issues; you will cause performance problems; and you will architect things poorly.

You cannot live in a vacuum: you have to talk to users and stakeholders and your teammates and your manager. But that should be done on your schedule, not on theirs (most of the time) so that when you are done talking to them and you have a good idea of what to build, you can go crank out a high quality first version. This version will be on the right track, technically: good architecture, usable performance, well-tested, with minimal bugs. This is a first iteration you can go take to users to get concrete feedback and keep iterating.

Our attention is being squandered and we have an opportunity now to reclaim it. Fight back. Get rid of the ping-pong table; delete Facebook and Snapchat; disable push notifications for emails; build some walls to establish real offices. Setup processes on your teams to give people large chunks of time where they can go deep, for days at a time. Put walls or even hundreds of miles between your employees. Embrace flow, and get some work done. You'll feel better, I promise, and what you produce will be better as well.

53 Avoid multitasking to write better code

Originally published on 2018.10.26.

Multitasking is incredibly alluring. Why go slowly, doing one thing at a time, if you could get a second thing done? Why not fill those five seconds while your code compiles with reading an article about the latest web frameworks?

In fact, multitasking is hiding everywhere in your daily work. Any time you switch from one task to another with the intention of going right back, that counts as multitasking. You might do it without realizing, because...

- while your code is compiling, you switch to your browser to check Twitter.
- while you are coding, you check Slack briefly to see if anything is going on.
- during a meeting, you check Hacker News for anything interesting.

The siren song of multitasking is strong, but the costs are high. Computers are designed for multitasking and parallelism, but humans are not. For us, context switching is very expensive: every time you switch, you lose track of where you were in the previous task, and it can take you 15 minutes to get the state of your codebase back into your head when you switch back. Is it worth spending **minutes** getting back to where you were just to save the mere **seconds** you wait for something else to happen? It certainly is not.

This constant context switching also drives down your quality. Every time you context switch, you drop ideas and you drop your focus and this means you cannot engage in very deep thought. You might be able to produce simple CRUD apps this way, but even then, you will miss subtleties in your data model or the domain you are solving for or in how your users will engage with the application. The quality of what you ship decreases when you work in short bursts.

Instead of working in bursts, the way our phones have trained us to do, it is better to spend that time simply idle or bored (in shame, I admit that I looked at Twitter three times while writing this post). Boredom is critical to getting good work done, because the downtime lets your brain explore the non-obvious paths that you may not go down otherwise. This is how you find major bugs, deficiencies in your architecture, or unexpected user experience issues before you ship it.

Multitasking will also make you ship your code late. When you provide estimates, those estimates are usually given optimistically. They are written with the assumption that the requirements are complete, no unexpected complexity is hiding in the problem, and most importantly, that your time will be allocated in sufficiently large chunks to the task. To see this is true, think of a programming task that would take you two hours to complete. How long would it take you if you can do it all in one shot? What about if you can only work in 30 minute chunks? What if you can only work in one minute chunks? If you are limited to one minute chunks and the task has any reasonable complexity, you might **never** finish it.

It is very rare that you truly need to multitask or switch between tasks rapidly. (If your job is one where you are expected to respond to instant messages instantly, I am truly sorry, and you need to know that there are better opportunities out there. I'm always happy to help point people toward better jobs.) In almost all of your daily work, you can afford to let things go. Those posts on Twitter will still be there when you are at a good stopping point. Those Slack messages will still be there when you are at a good stopping point with your code. That HackerNews post will still be there when your meeting is over. The task you are working on, or the peers you are working with, deserve your full attention, and giving it will let you ship higher quality software on a more reliable schedule.

54 Kill the crunch time heroics

Originally published on 2018.11.02.

Crunch time has an allure: it feels like if you just push hard enough, you can get more done. You can push hard and get that next release done on time, get those new features out, earn more revenue for your company. Engineers are under immense pressure to deliver more and do it now, and we also feel special: we feel unique, like we are not subject to the fatigue that others experience, or that this project is different and we can do it even when exhausted.

None of this is true, though. We are not special. We are all humans, and for all of us, crunch time is expensive. Some people literally die as a result of crunch time; most of us just end up as worn out shells in poor health, making poor products.

This happens because crunch time is inherently very fatiguing, physically and mentally. During crunch time, you don't get the sleep that your body needs to maintain itself and to recharge the brain, so you end up grouchy and tired and sloppy. During this time, you make big sacrifices: you see your kids less and your partner less; you give up some of your hobbies; you sacrifice your health by giving in to temptations and unhealthy food and alcohol. This all can go to extremes, as happened in 2013 at Bank of America, where an intern died after working for three straight days.

The costs to personal lives are bad enough, but you'd expect them to be justified with something really valuable to the business. I mean, why else would a business press people to do this to themselves? Unfortunately, it isn't so. Long-term, crunch time kills businesses. It leads to lower morale and disengaged employees. Quality declines and passes into negative returns. The mistakes people make while they are fatigued are incredibly expensive, dropping databases or crashing cars or opening gaping security holes.

Look, I know these costs. In the face of these, I decided recently to do some crunch time. We were working on a plan for a prototype and there was a board meeting coming up, and my teammate and I decided to go for it: "let's try to ship a prototype ahead of the board meeting!" It was an ambition born out of a threefold desire: to elevate our company at the board meeting; to elevate the reputation of our engineering team internally; and honestly, to make ourselves look *damn* good by shipping something incredible in a short timeframe.

We did it. We worked overtime and gave up our hobbies and time with our partners so that we could write a lot of code that ended up looking really slick, working pretty well, and impressing our audience. I actually feel a lot closer to my teammate now, because he and I went through some tough stuff together and really got in sync.

But was it worth it? Definitely not. Looking back, I think that with some extra care (people shielding us from meetings; turning off Slack for us for that time; etc.) we could have gotten the prototype out on the same schedule without the overtime. Even if we missed that deadline, the days after that crunch time were essentially sick days for us both (we could not think clearly enough to write good code) and the week after it was not a great week, either. And the sacrifice to ourselves, to our partners, to our cats - that was something we took too lightly.

The software industry glorifies crunch time. We have this hero mythos where the cowboy coder goes into a cave with a bottle of Mountain Dew and emerges, a sleepless night later, with a beautiful, functional prototype. It's time to kill this hero. This is not how humans work: humans need sleep; humans need downtime to recharge; and humans deserve time for their friends, family, and hobbies.

Let's all fight back against crunch time. You deserve your own time and your own energy, and your business and product will be better for it. Incentives are aligned here: everyone wins when you avoid long hours and crunch time. Let's say no, together.

55 Books I Read in 2018

Originally published on 2018.12.31.

Every year, GoodReads[1] has a Reading Challenge, where you set how many books you want to read and record them as you go. This year, I got serious about it, and it was a wonderful motivational device. I set a goal of two books per month, and I just eked it out over the finish line, finishing my 24th book this morning.

Here are some of the best.

55.0.1 Rework[2] and Remote[3]

These are two modern classics by DHH and Jason Fried. These fit so well into my thoughts about what a workplace should be and the culture we should cultivate that they weren't mind blowing - rather, they were an incredible distillation of things I've *wanted* to say, in an incredibly clear manner. These two alone let me point to them and say: this is the kind of company I want to build.

Don't worry, the third in this series, It Doesn't Have To Be Crazy at Work[4], is one of my first books to read in 2019.

55.0.2 Salt, Fat, Acid, Heat[5]

This book has changed the way I cook. It teaches you the fundamentals - mentioned in the title - and how to understand and apply them to any dish you are cooking. I was a good cook before this, but now I'm a vastly more capable home chef.

For anyone looking to step up their cooking game, to really understand what they are doing and break their reliance on recipes, this is a fundamental you deserve to have on your shelf.

---

[1] https://www.goodreads.com
[2] https://www.goodreads.com/book/show/6732019-rework
[3] https://www.goodreads.com/book/show/17316682-remote
[4] https://www.goodreads.com/book/show/38900866-it-doesn-t-have-to-be-crazy-at-work?ac=1&from_search=true
[5] https://www.goodreads.com/book/show/30753841-salt-fat-acid-heat

55.0.3 Built: The Hidden Stories Behind our Structures[6]

Have you ever wondered how subway tunnels are dug under rivers? How skyscrapers are built to withstand disasters? How structures stay standing for so many years? Well, this is the book for you. This is an incredible peek into what actually goes into creating and maintaining the structures we rely on every day to live in, drive through, and work from. Hands down one of my favorite books I've read all year.

55.0.4 The Three-Body Problem[7] and The Dark Forest[8]

Such an excellent series. It is shockingly well written, and the credit is due to both the author and the translator, who is himself an award-winning sci-fi author. This series is one of the best I've read in a while. It is both an interesting universe and a believable one, with good characters and interesting plot.

55.0.5 The Monk of Mokha[9]

This is a fascinating true story behind a young man's attempts to bring Yemen's coffee to the American market. Whether or not you are interested in coffee, this is a fascinating story which shows you the human side of the production of this little brown bean.

55.0.6 And the rest...

The rest of the books were good, but you can only have so many favorites. These are presented in reverse chronological order of my reading:

- A Hologram for the King[10]: a fantastic novel by the great Dave Eggers.
- Never Split the Difference[11]: an interesting perspective on negotiation with many intensely interesting anecdotes. I'm not sure how much of this I really can apply.

---

[6] https://www.goodreads.com/book/show/34921647-built
[7] https://www.goodreads.com/book/show/20518872-the-three-body-problem
[8] https://www.goodreads.com/book/show/23168817-the-dark-forest
[9] https://www.goodreads.com/book/show/35215524-the-monk-of-mokha
[10] https://www.goodreads.com/book/show/13722902-a-hologram-for-the-king
[11] https://www.goodreads.com/book/show/26156469-never-split-the-difference

- The Checklist Manifesto[12]: this book makes me want to go make checklists for everything. I want to apply what I've learned here to software engineering, but that's going to be a slow process.
- High Output Management[13]: this book isn't just for managers. He defines "middle manager" in an interesting way which includes many (if not all) knowledge workers and it is very transferable into my daily work. Even just knowing that "dual reporting" is a thing - and how to manage it - is very helpful.
- Code Girls[14]: let's just say, this is an incredible peek into the work of these incredible women who helped the Allies win WWII. This should be part of the history curriculum.
- Nino and Me[15]: a fascinating story of the friendship of two legal scholars. This isn't the first Garner book I bought (Black's Law Dictionary was), nor is it the last (Garner's Modern English Usage), but it is certainly the most page-turning.
- Start Small, Stay Small[16]: music to my ears. This is an excellent resource on how to build a bootstrapped company, and something I intend to revisit.
- Coders at Work[17]: interesting interviews with some of the legends of software engineering and computer science. You must take it with a handful of salt. The biggest thing I took away was that all these successful people work in *incredibly* different ways, so there really is no single best way of working.
- Factfulness[18]: imagine a TED talk in book form, and this is what you get. Pretty interesting and eye opening.
- Skin in the Game[19]: this book got me to think about things from a perspective I have really never had before. I loved being challenged to think so differently. I'll read more Taleb. I won't buy everything he says, but it's worth reading for the new perspective.
- The Signal and the Noise[20]: such a good read by Nate Silver. Highly recommended.
- Hit Refresh[21]: this book made me realize how much Microsoft had changed under Satya Nadella's leadership. To an extent it feels like (and is) a marketing piece for Microsoft, but it boosts my confidence that some of their acquisitions (like GitHub) will live on and keep being wonderful.

---

[12] https://www.goodreads.com/book/show/6667514-the-checklist-manifesto
[13] https://www.goodreads.com/book/show/324750.High_Output_Management
[14] https://www.goodreads.com/book/show/34184307-code-girls
[15] https://www.goodreads.com/book/show/35566766-nino-and-me
[16] https://www.goodreads.com/book/show/9167158-start-small-stay-small
[17] https://www.goodreads.com/book/show/6713575-coders-at-work
[18] https://www.goodreads.com/book/show/34890015-factfulness
[19] https://www.goodreads.com/book/show/36064445-skin-in-the-game
[20] https://www.goodreads.com/book/show/13588394-the-signal-and-the-noise
[21] https://www.goodreads.com/book/show/30835567-hit-refresh

- The Big Short: Inside the Doomsday Machine[22]: a great look at what led to the housing bubble collapse and market crash of 2007-2008. Not really an uplifting topic, but a great read.
- A Wrinkle In Time[23]: this is a classic, and I'm glad to have finally read it.
- Do What You Love and Other Lies About Success and Happiness[24]: I love the title but did not enjoy the book. It was written in an incredibly dry, overly academic style.
- Fire and Fury[25]: Exactly what you'd expect from a book about the current administration written this early.
- Ready Player One[26]: this was a great look at how bad our future could be if we blindly lean into technology and corporatism. Let's not, okay?

2019 is going to be great, and I have a massive list of books I want to read (and a smaller list of ones I actually *will* read, as always).

---

[22] https://www.goodreads.com/book/show/26889576-the-big-short

[23] https://www.goodreads.com/book/show/33574273-a-wrinkle-in-time

[24] https://www.goodreads.com/book/show/23492333-do-what-you-love-and-other-lies-about-success-and-happiness

[25] https://www.goodreads.com/book/show/36595101-fire-and-fury

[26] https://www.goodreads.com/book/show/9969571-ready-player-one

56 Gmail's "Smart Compose" feature should be considered harmful

Originally published on 2019.02.27.

In 2005, I got my invite to get a Gmail account. It was incredible, and I loved it, although I didn't really know *why* at the time. It was a combination of really great design so it was pleasant to use, the hype built up by the invite system, the perpetual feeling of getting something more as you watched your allotted storage slowly tick up, and quite a bit from the fact that it was the first email account I signed up for on my own. I had an email account before that, created by my parents through our ISP, but this one was *mine*, created by me, from an invite my friend gave me, and all my friends were also using Gmail - if they could get an invite. With that ability to also *chat* through your webmail client... it was mindblowing, and it eventually supplanted AOL Instant Messenger for my friends and me.

For the last 13 years, Gmail has been my primary email service. (I recently switched to a paid provider who respects my privacy. I'm willing to pay for privacy and security.) It has been instrumental in allowing me to communicate and stay in touch with family, friends, and acquaintances as we spread out across the globe. I have 65,000 emails in my account, 10,000 sent emails, and 9,000 chats with friends.

Along the way, they added features that helped me communicate more effectively. When they added the Google Labs feature to stop an email from sending if you say "find attached" or something and forget to attach a document... that saved me embarrassment many times, and it saved my contacts from frustration. When they added video calls in 2008, it made it easy to actually see my parents while I was away at college. Filtering gives us the ability to sort incoming messages and reduce information overload. Labels replaced the folders of other mail providers as a more natural fit for how we want to organize our information. The superior search features of Gmail have made it so that I have a (slow, asynchronous) external memory where I can look up past events. And not least of all, Gmail's spam filtering was incredibly effective at a time when most mail providers struggled against the tide.

This is all to say: Gmail, I have loved you, but "Smart Compose" and "Smart Reply" are strong deviations from the true value of Gmail.

The true value of Gmail has been in enhancing and facilitating communication, especially sincere communication. "Smart Reply" hinders this value by reducing the overall variety of responses, and "Smart Compose" biases the words that you send.

With "Smart Reply", Gmail shows you three possible replies below your email, encouraging you to select one to reply "efficiently". The problem is that this has the same effect as naming a number in a negotiation: it anchors you around those possibilities and reduces your creativity in responding. If someone asks you whether Monday or Tuesday works for you, it will offer the responses "Let's do Monday," "Monday works for me," or "Either day works for me" (this is an actual example[1] from the announcement blog post). The net effect of this, I believe, is that people will be *less likely* to schedule on Tuesday, because the convenient options were all presented saying that Monday is better or that they are equal. For setting up a coffee date, this might not be a big deal, but what about for price negotiations? The models for "Smart Reply" are surely trained on real emails, so what if the model learns that men will negotiate more aggressively than women? If that makes its way into the "Smart Reply" feature, it will have material harm for any woman negotiating pricing over email by slightly biasing them toward less aggressive negotiations, reinforcing the status quo.

Similarl harms are baked into "Smart Compose", a feature which suggests the next word or phrase for you to type based on what you've typed so far. They've already had to remove pronouns[2] because the system was biased toward men, so it is difficult to believe that it is unbiased in all other ways. What other harms are in the system that Google engineers simply have not detected yet?

And that's all just the active immediate harm from a particular message. There is also the more subtle shift from automating some of our communcations. What if the black box learns when your contacts' birthdays are, and suggest sending "happy birthday" to them on those days and fills it in for you? How long will it take to erode the well-wishing tradition in our society, replacing it with a mechanical button-press? The point of saying "happy birthday" isn't to just say it - it is to actually think about them and take your time to call them or message them with your well wishes.

Automating our communication is done in the name of efficiency but it is robbing us of the one thing that makes humans human: our language and our communication. It is causing direct harms, whether through something as benign as a "happy birthday" or through something as sinister as biasing negotiations or oppressing whole groups. I hope these harms

---

[1] https://www.blog.google/products/gmail/save-time-with-smart-reply-in-gmail/
[2] https://www.reuters.com/article/us-alphabet-google-ai-gender/fearful-of-bias-google-blocks-gender-based-pronouns-from-new-ai-tool-idUSKCN1NW0EF

were considered by the product managers at Google, but these are public harms, so the users and the public deserve to be made aware of whatever tradeoffs have been made and whatever protections are in place. In the meantime, we should all consider these features harmful and a net negative for our society.

57 Terminology matters: let's stop calling it a "sprint"

Originally published on 2020.04.29.

If you're in the software industry, it's hard to not be aware of agile development at this point. It seems like every team practices it differently, but there are certain commonalities that run through all teams I've seen. One of those is the term used for each time-delimited section of the development process: "sprint."

I'm an endurance athlete, and this term sends shudders through me. Software development is very much akin to an endurance event. You run into similar challenges. When you're running a marathon, most of the work is already done, if you have trained adequately, but there is a lot remaining still to do during the event itself. It's a mental game at that point: you need to have the resolve to just keep putting one foot in front of the other, over and over, over and over, until you hit that finish line hours later. But here's the thing: at no point during a marathon do you – or should you – sprint. Sprinting is high effort and high speed and can be sustained for some time, but not for 26.2 miles. If you sprint at any point during the race, then you are decreasing your overall performance, because that spent energy reduces the capacity you have to run the rest of it at your max sustainable pace.

Software development is similar. Our brains are not infinite resources which we can push day-in and day-out. This is why we have to sleep, so our brains and bodies can recover from the toils of the day. It is well known that as we work longer hours, our output gets slower and slower, and that it is able to reach negative returns - so there is a point where working longer hours reduces your total output. It actually does not take very long to reach that point.

That is what is actually analogous to a sprint: something which is so taxing for you to do that it reduces your capacity for other exertion temporarily, which you need to put significant effort toward recovering from. A normal development cycle is not, or should not, be a sprint, because you have to do many of these in a year, over and over, without an end in sight. Even if you leave that team, you will wind up somewhere else where you are repeating these development cycles. It would be better that we call them something else: perhaps a "leg", continuing the running analogy but this time evoking a long journey ("leg of the journey") or relay race.

This may seem inconsequential semantic nitpicking. It is not. The terms we use set expectations for those inside and outside our industry. If you have little other context around how software development works (if you're new to the industry, if you're hearing a relative talk about work) then when you hear "sprint" it will make you think of a high exertion activity, such as you put in at crunch time when you need to just push something over the finish line. Even within our teams, the term shifts mindsets and can justify problems. Most of us have probably been in situations where we justified doing things in a very hacky way since it's temporary, just to shove something forward, "we'll fix it later" (we never do). This is undoubtedly influenced by the terminology we use. Every day we hear the term "sprint." "Next sprint, we'll fix that," we say. "This sprint, we're doing it quickly."

This would all be different - subtly, but surely - if we used a more fitting term. If we called each development cycle a "leg", it could evoke many images but in the context of a journey, would surely shift our mindsets to think more about how this is really just part of our longer journey to create a product, build some features, change the world. It puts the emphasis on this cycle being part of a larger whole. That will change what you and your teams produce, because it raises your consciousness of the long-term impact of what you do.

58 Parallel assignment: a Python idiom cleverly optimized

Originally published on 2020.05.15.

Every programmer has had to swap variables. It's common in real programs and it's a frequently used example when people want to show off just how nice and simple Python is. Python handles this case very nicely and efficiently. But *how* Python handles it efficiently is not always clear, so we'll have to dive into how the runtime works and disassemble some code to see what's happening. For this post, we're going to focus on CPython. This will probably be handled differently in every runtime, so PyPy and Jython will have different behavior, and probably will have similarly cool things going on!

Before we dive into disassembling some Python code (which isn't scary, I promise), let's make sure we're on the same page of what we're talking about. Here's the common example of how you would do it in a language that's Not As Great As Python:

```
# assume we have variables x, y which we want to swap
temp = x
x = y
y = temp
```

Okay, so we've all seen that, what's the point, I'm closing the tab now. Well now we get to the part that's people trumpet as evidence of Python's great brevity. Look, Python can do it in one line!

```
# assume we have variables x, y which we want to swap
x, y = y, x
```

This method is known as parallel assignment[1], and is present in languages like Ruby, as well. This method lets you avoid a few lines of code while improving readability, because now we can quickly see that we're doing a swap, rather than having to look through the lines carefully to ensure the swap is ordered correctly. And, we might even save some memory, depending on how this is implemented! If you followed

---

[1]https://en.wikipedia.org/wiki/Swap_(computer_programming)#Parallel_assignment

the link to the Wikipedia article about parallel assignment, you'll see the following:

> This is shorthand for an operation involving an intermediate data structure: in Python, a tuple; in Ruby, an array.

A very similar statement is made in Effective Python[2] (a great book to read together as a team, by the way!), where the author states that a tuple is made for the right-hand side, then unpacked into the left-hand side.

This makes sense, but it isn't the whole story, which gets **far** more fascinating. But first, we need to know a little about how the Python runtime works.

Inside the Python runtime (remember that we're talking about CPython specifically, not Python-the-spec), there's a virtual machine and the runtime compiles code into bytecode which is then run on that virtual machine. Python ships with a disassembler[3] you can use, and it provides handy documentation listing all the available bytecode instructions[4]. Another thing to note is that Python's VM is stack based. That means that instead of having fixed registers, it simply has a memory stack. Each time you load a variable, it pushes onto the stack; and you can pop off the stack. Now, let's use the disassembler to take a look at how Python is *actually* handling this swapping business!

First, let's disassemble the "standard" swap. We define this inside a function, because we have to pass a module or a function into the disassembler.

```
def swap():
  temp = x
  x = y
  y = temp
```

This doesn't do anything useful, because it just swaps them in place. We didn't even declare the variables anywhere, so this has no chance of ever actually running. But, because Python is a beautiful language, we can go ahead and disassemble this anyway! If you've defined that in your Python session, you can then `import dis` and go ahead and disassemble it:

---

[2]https://effectivepython.com/
[3]https://docs.python.org/3/library/dis.html
[4]https://docs.python.org/3/library/dis.html#python-bytecode-instructions

```
>>> dis.dis(swap)
  2           0 LOAD_FAST            0 (x)
              2 STORE_FAST           1 (temp)

  3           4 LOAD_FAST            2 (y)
              6 STORE_FAST           0 (x)

  4           8 LOAD_FAST            1 (temp)
             10 STORE_FAST           2 (y)
             12 LOAD_CONST           0 (None)
             14 RETURN_VALUE
```

Stepping through this, you can see that first we have a LOAD_FAST of x which puts x onto the top of the stack. Then STORE_FAST pops the top of the stack and stores it into temp. This general pattern repeats three times, once per line of the swap. Then, at the end, we load in the return value (None) and return it. Okay, so this is about what we'd expect. Barring some really fancy compiler tricks, this is analogous to what I'd expect in any compiled language.

So let's take a look at the version that is idiomatic.

```
def swap():
  x, y = y, x
```

Once again, this isn't doing anything useful, and Python miraculously lets us disassemble this thing that would never even run. Let's see what we get this time:

```
>>> dis.dis(swap)
  2           0 LOAD_FAST            0 (y)
              2 LOAD_FAST            1 (x)
              4 ROT_TWO
              6 STORE_FAST           1 (x)
              8 STORE_FAST           0 (y)
             10 LOAD_CONST           0 (None)
             12 RETURN_VALUE
```

And here is where we see the magic. First, we LOAD_FAST twice onto the stack. If we just go off the language spec, we'd now expect to form an intermediate tuple (the BUILD_TUPLE command is what does this, and from its absence we know that we aren't building a tuple here the way you would with x = (1,2)). On the contrary, you see... ROT_TWO! This is a cool instruction which takes the top two elements of the stack and "rotates" them (a math term for changing the order of things, kind of shifting everyone along with one moving from the front to the back). Then we STORE_FAST again, twice, to put it back into the variables.

The question now might be, "why do we even need ROT_TWO? Why can't we simply change the order we store them to achieve the same effect?" This is because of how Python has defined its semantics[5]. In Python, variables on the lefthand side of an expression are stored in order from left to right. The righthand side also has these left-to-right semantics. This matters in case like assigning to both an index and a list:

```
a = [0, 0]
i = 0
i, a[i] = 1, 10
```

If you didn't define the semantics, the result above would be ambiguous: will a be [0, 10] or [10, 0] after running this? It will be [0, 10] because we assign from left to right. Similar semantics apply on the righthand side for the comma operator, and the end result is that we have to do something in the middle to ensure we adhere to these semantics by changing the order of the stack.

So, at the end of the day, there you have it. Parallel assignment, or swapping without another variable, does not use any extra honest-to-goodness tuples or anything under the hood in Python. It does it through a clever optimization with rotating the top of the stack!

Update 5/16: I made a few edits to make the article clearer and avoid distracting from the content by implying/stating that people were wrong, and making certain things clearer (focus on CPython, focus on implementation vs. spec, etc.).

---

[5] https://docs.python.org/3/reference/simple_stmts.html#assignment-statements

59 Where is the source code for ping?

Originally published on 2020.07.26.

Lately, I've been working on implementing `ping` on my own as a project to keep learning Rust and to deepen my knowledge of networks. I'm just going for a super basic utility here, nothing fancy, not even all the features of `ping`. But since the language is new to me *and* my lower-level network knowledge is weak, I decided that it could be helpful to compare notes, so to speak, with the real deal itself. So that's our question: where is the source code for the actual utility `ping`?

Let's find out! I'm running Ubuntu, so the question is where do my binary packages come from and where does the corresponding source live? Naively, I expected this to be super easy to find. It's not *hard* to find, for my system (Ubuntu), but it's not as easy as it would be on some others like Gentoo.

The first step I took was, naturally, to turn to Google and search "ping source code". The first search result is a GitHub gist[1], which links to a US Army page[2] written by the original author of ping. Cool, so this is seems like the original source! This is really cool and a great historical artifact. Is this the same version that I'm running on my desktop, though? We need to dig deeper and see what's running on the local machine.

If we use `man ping` to look at the manual page for ping, we see "System Manager's Manual: iputils" at the top, which is our first hint at where ping comes from on our system: possibly the package is named iputils, and I do have a package named iputils-ping installed. From here, we can find the source package[3] and... the links on that page to the Debian git repos are broken. Sigh.

Back we go to Google and we find the source package for iputils[4] on Debian, figuring it's probably the same. And now we're in luck, and we can get to the iputils git repo that Ubuntu presumably draws from by way of Debian: https://salsa.debian.org/debian/iputils

---

[1] https://gist.github.com/kbaribeau/4495181
[2] https://ftp.arl.army.mil/~mike/ping.html
[3] https://packages.ubuntu.com/source/bionic/iputils
[4] https://packages.debian.org/source/buster/iputils

And thus we find the source: https://salsa.debian.org/debian/iputils/-/blob/master/ping.c and https://salsa.debian.org/debian/iputils/-/blob/master/ping.h

It clocks in at a total of approximately 2k lines of code (well, there is also shared code), which is not much bigger than the original source. It's a marvel to me that this gem of software has stayed small, concise, and *useful* for decades without acquiring much bloat, without changing forms. May more software be like ping.

Now that we have the source, there's a lot more to learn. For example, if you receive pings intended for another process (because that happens with raw sockets, it turns out), you can setup a filter with Berkeley Packet Filter, and ignore any pings that aren't for you! This is really cool and something that I need to learn more about.

What other gems am I missing out there?

60 What's "good" code and does it matter?

Originally published on 2020.10.14.

I take pride in my work and in writing good code, and it's important sometimes to take a step back and ask: what does that even mean? And does it matter?

At a high level, "good code" is code that is suitable for its purpose and achieves its goals. That definition is pretty lacking, though, I think. You can write some very very hacky prototypes that achieve their goals—proving out an idea—while also being pretty objectively bad code. But objectively, by what measure?

This is where we get back into what it means for code to really be suitable for its purpose. Code is a living thing. It is written, edited, read, and used, in order of increasing frequency: most code will be used far more often than it is edited, read far more often than it is edited, and edited far more often than it's newly written. This means that for code to be good, it has to support these activities. It has to do its job well when it's used. It has to be able to be read. It has to be able to be updated.

A great deal has been written already on how code can do its job well. The short summary here is that it has to do what is expected (per the spec, if you are fortunate enough to have one) and have few defects. On the non-functional side, it has to also do what's expected in a reasonable amount of time, reliably. It doesn't matter how free of bugs your program is if it literally never terminates. And sometimes a "reasonable amount of time" might actually be a floor on the time for things like bcrypt, which we *want* to make reasonably slow.

Supporting reading and editing go hand-in-hand, because they are core parts of maintaining a codebase. You cannot really edit a codebase confidently if you cannot read it and understand what it's doing, and you cannot fix bugs or add features if you cannot edit it confidently. While tests are a big portion of this, they are distinct from the quality of the code under test. In an ideal world, they add assurance, but the code itself should have a clear design that presents itself. It should be designed from the outset to be extensible.

But does that really capture what we're doing day-to-day? I'm a software engineer, not a computer programmer. While I take pride in writing good code, my job is not to produce good code but to effectively solve problems, usually using code. In practice, engineering means you have to make tradeoffs.

When you're trying to solve a problem, but you're not sure exactly how to solve it, you reach for prototypes and proofs of concept. These will be sufficient to test an idea and validate the approach, but you can cut a lot of corners on them. The code doesn't look good, it's almost certainly not maintainable long-term. But is this a good engineering decision? In a lot of cases, yes! It's the right tradeoff to make.

Similarly, you can write the absolute best code you have ever created for that shiny new feature, but... realistically, you're probably working on it in a business, and realistically, improving that quality to make it super readable and super extensible won't deliver value to the business. It really depends on how much the code will be extended and read, and it's also a tradeoff between time now (for a startup burning cash, time right *now* is in very short supply!) and time later (once you get profitable or take another infusion of cash from ~~rich suckers~~ venture capitalists, you can afford to rewrite things).

This comes back in a lot of decisions you have to make as a software engineer. If you design a super-scalable system that can handle all the traffic you will need three years from now... well, that growth will probably never materialize, because you did not spend that time developing your product right *now*. It's often a better decision to write something that works okay for now, and refactor/rewrite later when you need to scale up.

So, does writing good code matter?

It does, to an extent. Your code has to be good enough to do its job, which is usually to deliver value and to optimize more for the here and now than down the road, scale that might never materialize. If your code does its job well enough and it can be (maybe somewhat painfully) maintained and updated for a couple of years, well, by the time those two years are gone you may well have rewritten it anyway! If you'd spent twice as long at the outset writing your magnus opus, that time would have been wasted.

Your code can't be a dumpster fire. It probably shouldn't be the Mona Lisa, either. Striking a balance is an important facet of engineering for all things, ranging from the quality of your code (good, but not *too* good!) to how much scale to handle (enough, but don't over-engineer it!) to how much coffee to drink (just kidding, never too much coffee).

.

61 Solving my fun, frustrating docker-machine error

Originally published on 2020.12.08.

Last Saturday, I ran into a problem doing a routine backup of a web app I maintain. In fact, this was the *second* time I ran into the *exact* same issue, so it's time to write it down. (Hopefully, the third time I run into this, I have the presence of mind to look up my own solution!)

My web app is deployed using docker-machine and docker-compose. This is not a great production setup, but it works for me and there are just a handful of users. Every week, I manually run a backup script that copies down the database and all the images from this web app. (I could set up a cron job, but I consciously chose to keep it manual so I would, every week, be able to see that the backups are working: this has paid off, since I saw when it did NOT work!)

When I ran the backups, I ran into a mysterious error message:

```
$ backup.sh
Error checking TLS connection: Error checking and/or regenerating
↪  the certs: There was an error validating certificates for host
↪  "xx.xx.xx.xx:2376": dial tcp xx.xx.xx.xx:2376: i/o timeout
You can attempt to regenerate them using 'docker-machine
↪  regenerate-certs [name]'.
Be advised that this will trigger a Docker daemon restart which
↪  might stop running containers.

Error response from daemon: Container xxx is not running
```

First thought: Okay, well, that's weird that the certs are expired but let's just follow what it says. Let's regenerate those. So, I did, and then... the entire app was down, because it shut down the containers but could not start them! Now a routine backup has turned into an outage.

Aaand I can't see the machine:

```
$ docker-machine ls
NAME            ACTIVE   DRIVER    STATE     URL    SWARM    DOCKER
↪  ERRORS
picklejar                generic   Timeout
```

The strange thing? `docker-machine ssh <host>` worked.

So... I cannot see the machine, I cannot validate the certs, but ssh works.

If you're screaming at the monitor right now because the answer is obvious, *I know*. I missed it in the moment, but it was right there in front of me (sort of) in that first error message: `dial tcp xx.xx.xx.xx:2376: i/o timout`: This means that we can't establish a TCP connection on that port, which could be... caused by the firewall. Let's not talk about how long it took me to realize this, and how many other things I tried before I had that head smack moment: doh!

The problem was: I have the instance firewalled in a way that allows my home network to establish the TCP connections needed for docker-machine, but no external traffic. BUT I have ssh allowed from *any* port, so that I can get into the host while I'm on the go (or that was the idea, when travel was a thing). So when my ISP issued me a new IP address, suddenly I could do some things on the machine (ssh) but could not do others, leading to this confusing situation of docker-machine kinda sorta half working.

So if you get an error message from docker-machine about an error validating certificates, don't just assume (as I did) that its suggested fix is a good idea: verify that you don't have a network/firewall issue first.

62 Tech salaries probably aren't dropping from remote work

Originally published on 2020.12.22.

Not even a year ago, most software companies and software engineers were some form of remote work skeptical. Remote work existed (I've been working remote for most of my admittedly short career!) but it was not widespread. When I talked to recruiters at big tech companies they would all insist that remote work was not feasible for them, and even at the companies I worked for, there was pushback that this definitely wouldn't work for us because *reasons*. But now, I think we're seeing one of the real reasons people were skeptical:

Money.

There are lots of very real reasons to like colocation, and there is a lot that's hard about remote work. But I haven't seen a lot of discussion around the role of salaries in affecting whether or not people prefer remote work. Imagine this (and for some of you, this might be *really* easy): you're working in New York or San Francisco, pulling down an absurdly high salary. You know that in the Midwest, salaries are much lower, not even comparing to other countries where salaries might be *much* lower. In this situation, it seems pretty rational to oppose remote work. If you only allow colocation, then you're competing against other workers who all have the same cost of living as you and all have the same sky-high salary demands as you. But if you allow remote work, that great dev from Akron, Ohio might be willing to do the same job for less money. Right?

Well... it isn't quite that simple. Right now, tech salaries have been going up consistently for a while (anecdotally, I've seen this as long as I've been here, so at least since 2013). This is consistent with demand for software engineers outpacing supply of them. Companies like Basecamp and Stripe are employing software engineers *anywhere* at California rates. These companies aren't charities: they are doing it because they understand that, in the market conditions we have right now, to get the employees they need for their business to be successful, they have to pay those rates.

Now the pool of software engineers who are available for remote work has expanded dramatically. So has the competition, as many companies are leaning into this advantage in hiring (including my employer) by looking for the *best* talent, regardless of where it is. When you do that, you have to pay what the competition is willing to pay, and right now that's going to be... *checks watch*... a pretty high salary and good equity compensation.

The gut check on this is to consider outsourcing to other countries. This has been a trope and fear as long as I've been aware of computers. For most of my life, people have been talking about how software development is all going to go overseas to whatever the current country of interest is: India, Russia, Ukraine, China... you hear people saying these countries with cheaper labor are going to eat our industry and we'll be out of jobs. Well, it hasn't happened. So why not? There are a few advantages to hiring an engineer in the US: you get someone who is on a compatible timezone (you could also get this in South America), who shares the same language, who has shared cultural touchpoints. And you're working within the local legal framework, which is easy: hiring across US state lines is harder than hiring in the same state, and hiring out of the country is harder than hiring within the same country. So it's just far more convenient to work with people who are in the same market, usually. But we *are* hiring people globally, and many teams are distributed around the globe. The effect of that? It's buoying salaries everywhere. The countries which once had the cheapest labor have been getting wealthier and salaries have been rising, and before you know it, the labor arbitrare will be hardly worth it (if it even is now).

So you can rest easy, probably: your tech salary is safe with remote work.

But of course I'd say that, because I work remote and like my salary ☺. So make your own judgments, and take your and others' biases into account. From where I sit, it looks like remote work doesn't dramatically change the market forces that are keeping salaries up. I'd be more afraid of bootcamps (and maybe when those threaten our salaries, you'll see a sudden push for licensing to keep salaries up and add barriers to entry).